Just-in-Time Teaching

Blending Active Learning with Web Technology

Gregor M. Novak
Indiana University-Purdue University

Evelyn T. Patterson
United States Air Force Academy

Andrew D. Gavrin
Indiana University-Purdue University

Wolfgang Christian
Davidson College

Prentice Hall
Upper Saddle River, NJ 07458

Library of Congress Cataloging-in-Publication Data
Just-in-time-teaching : blending active learning with web technology /
Gregor M. Novak . . . [et al.].
p. cm.
Includes bibliographical references and index
ISBN 0-13-085034-9 (pbk.)
1. Physics--Study and teaching (Higher) 2. Internet (Computer network) in education. I. Novak, Gregor, M.
QC30.J85 1999 99-10512
530".078"54678--dc21 CIP

Executive Editor: Alison Reeves
Editorial Assistant: Gillian Kieff
Production Editor: Kim Dellas
Manufacturing Manager: Trudy Pisciotti
Copy Editor: Barbara Booth
Art Director: Jayne Conte
Cover Designer: DeFranco Design, Inc.

Upper Saddle River, NJ 07458

Printed in the United States of America

10 9 8 7 6 5 4 3 2 1

ISBN 0-13-085034-9

Prentice-Hall International (UK) Limited, *London*
Prentice-Hall of Australia Pty. Limited, *Sydney*
Prentice-Hall Canada Inc., *Toronto*
Prentice-Hall Hispanoamericana, S.A., *Mexico City*
Prentice-Hall of India Private Limited, *New Delhi*
Prentice-Hall of Japan, Inc., *Tokyo*
Simon & Schuster Asia Pte. Ltd., *Singapore*
Editora Prentice-Hall do Brasil, Ltda., *Rio de Janeiro*

To all our students

Contents

Foreword

by Jeanne Narum

Again and again, the community of academic physicists calls for rapid and widespread dissemination of "best practices" in undergraduate physics education—those that are firmly grounded in rigorous physics education research and that are recognized successes in strengthening student learning. The community also calls for dissemination mechanisms that include thoughtful reflections on the intellectual and pedagogical basis for such innovations and mechanisms to aid others in adapting those new approaches in the classroom and lab.

Just-in-Time Teaching: Blending Active Learning with Web Technology responds to that call. In this volume, educators Novak, Patterson, Gavrin, and Christian set forth a carefully reasoned and well-documented plan to "...humanize instruction for all [physics] students and make a real difference to the non-traditional student." As physics faculty in undergraduate settings look for ways to change the curriculum to appeal to a wider spectrum of beginning students and to make certain that physics remains an integral part of liberal arts education, the strategies outlined in this curricular guide will serve them well.

The goals established by the designers of JiTT (Just-in-Time Teaching) reflect attention to, and understanding of, some of the particular challenges facing educators in the physics community—lack of student motivation and preparation for the rigor of learning physics, accompanied by a lack of appreciation of the value and utility of the kind of skills acquired in introductory physics classes and labs. Their intentions are unequivocal: to get students motivated by connecting what they do in the classroom to the world in which they live and work; to help students gain a deeper understanding of the world of physics through a series of interactive lectures, guided discussions, and Web-based preparatory work; and to help students realize that what they are learning will help them tackle problems in other courses and give them a "can do" attitude that will serve them well in all types of careers.

These goals and intents mirror those pursued by faculty in other SME&T disciplines who share the dream of transforming the learning environment for their students—for *all* of their students. Though not a physicist myself, I have observed many demonstrations of JiTT at Project Kaleidoscope workshops, and my response is always: *"Of course. What a good idea! Why haven't others thought of this?"* In a time when educators at the faculty and the administrative level are seeking to make the best use of technology in the service of learning, when physicists on campuses across the country are looking for approaches to undergraduate physics that attract students to introductory classes and bring them to understand that physics is a dynamic field with important applications to understanding both the natural world and technology, the lessons learned from the experience of JiTT should be welcomed by the community.

> I distinguish two aspects of science: content and enterprise; one is classified knowledge, the other is the way in which scientists work and think. The one way in which we write up our results, in papers and books, in the passive voice, gives the impression that we start with precise measurements and proceed by strict logical steps to incontrovertible conclusions. The way we really do it—starting with hunches, making guesses, making many mistakes, going off on blind roads before hitting on one that seems to be going in the right direction—that is science in the making.
>
> —Berkeley chemist Joel Hildebrand, 1957

One lesson learned from reviewing this text and observing the work of its authors is the importance of space. In discussions about "technology for learning," too often we hear that *space* will no longer be needed, that as faculty and students are able to communicate electronically and meet virtually, they will no longer need actually to come together.

The approach of the developers of JiTT reaffirms the importance of a face-to-face community of learners as a part of the total learning experience. The interactive lectures and collaborative recitation exercises are central to the success of JiTT because they bring students, mentors, and faculty together into a community that works collaboratively in addressing unfamiliar problems. The engagements (student-student, student-mentor, student-faculty) within the classroom help to underscore the point that physics is a human activity, that collaborating in problem-solving is how scientists "do science." Given that, however, this approach also underscores the usefulness of Web technology as a teaching tool when it is thoroughly blended into the total program.

> When we speak of the classroom as a science community, we picture an organization based on dialogue and activity. Knowledge is not transmitted so much as it is constructed, cooperatively, by students working together under the guidance of faculty and—at more advanced levels—by students and faculty working as teammates.
>
> —Project Kaleidoscope, Volume I, 1991

One question that persistently surfaces in considerations of curricular reform is: *How do you know it works?* As with many emerging reforms, we find that greater attention is paid to that question in the JiTT approach than in more traditional methods. Their goals are explicit and well articulated. They want students to take charge of their own learning; gain the skills of problem solving, collaborating and communicating; and have a conceptual understanding of physics. From these goals, the JiTT team draws a direct line from the activities they have designed to the mechanisms they use to evaluate the impact on student learning. Those departments that are hesitating to explore new pedagogical approaches and teaching tools will find in this text a carefully reasoned argument about methods that succeed in making physics more exciting and real to students.

Perhaps the most crucial lesson to be learned from the experiences of the JiTT team is how important it is to know the students: to listen to and understand their past educational experiences, to identify the road bumps of learning physics for the individual student as well as for the class as a whole, and to consider where students go when they move forward from your course. To know the students this well, and to give them responsibility for their own learning (student-active learning, discovery-based learning, Just-in-Time Teaching, etc.) requires a special kind of attitude and approach from the faculty.

Again and again I hear comments from physics faculty about the need to change the departmental image as a step toward increasing enrollments. Coupled to these comments are questions about how to engage the interests and enthusiasms of this generation of students, about how best to transfer our knowledge to the students so that they will succeed, about how to make physics relevant to the students of today. Faculty serious about pursuing answers to those questions will find this curricular guide to be a valuable resource.

To a significant extent, physicists were among the pioneers in this focus on students, in undertaking systematic research on how students learned and understood physical concepts, and in integrating what they observed into the learning environment for their students.

> In the matter of physics, the first lessons should contain nothing but what is experimental and interesting to see. A pretty experiment is in itself more valuable than 20 formulae.
>
> – Albert Einstein

In this day, concerns about "first lessons" are common in all disciplines. On campuses across the country there are biologists and chemists, as well as mathematicians and Earth and planetary scientists, each working to transform the learning environment for their students. The Just-in-Time Teaching approach serves all students, no matter what their background, field of study, or career aspirations. The work of Novak, Patterson, Gavrin, and Christian may serve as a much needed bridge to bring the larger community together around issues of common interest and concern.

Foreword

by Alan Van Heuvelen

The twentieth century has seen many attempts at education reform. Most initiatives responded to crises such as Sputnik, economic difficulties, or war. The emergency efforts were short-lived. However, there have been successes. Over 600 studies in the last 90 years have compared classrooms using lecture-based instruction with classrooms emphasizing group work and cooperative learning. In a review of 51 high-quality studies, Johnson, et al., (1981) found that students in classes emphasizing cooperative learning and group work scored 0.88 standard deviations higher (almost one grade point) than matched students taught in a lecture-based instructional system.*

Many recent efforts have emphasized the need for more active participation by students in helping to acquire knowledge and in learning to use this knowledge for productive purposes. Richard Hake's report (1998) involving 6542 physics students found that a variety of active engagement methods produced an average 51 percent increase in the possible gain on a conceptual physics test compared to a 24 percent increase for students taught in lecture-based instruction.† The scores on a problem-solving test were also somewhat higher for the active-engagement students than for the lecture students. In these active-engagement methods that improve learning, students work with peers to make predictions, help construct knowledge, and solve regular and more complex problems.

In the '70s, the Personalized System of Instruction (PSI) became popular. Learning was divided into small units with well-defined objectives. Students studied on their own or with the help of undergraduate peer tutors. When ready, a student took a test on the unit and received immediate feedback about the outcome of that test. If the score was satisfactory, the student moved on to the next unit. If unsatisfactory, more study was required. Comparisons between PSI students and students taught in a lecture format indicated that the PSI system produced superior academic achievement. Hedges (1978) even found that PSI students were more successful than peer students from lecture sections in more advanced courses in

* D. W. Johnson, G. Maruyama, R. T. Johnson, D. Nelson, and L. Skon, *Psychological Bulletin* 89, 429–445 (1981).
† Richard R. Hake, *American Journal of Physics* 66, 64–74 (1998). Larry Hedges, *American Journal of Physics* 46, 207-210 (1978).

their own discipline and in science courses in other disciplines. It seemed that the PSI students had learned better how to learn.*

The Just-in-Time Teaching (JiTT) method described in this book takes advantage of many of the lessons learned from these previous successful teaching and learning innovations and adds promising new technology features.

- JiTT allows professors to adapt their lectures to student learning difficulties exhibited in electronic responses to questions and problems submitted before the lecture.
- Collaborative recitations provide individual interactions among students, tutors, and professors—a combination of the PSI tutor method and collaborative group work.
- Students use an on-line homework system that provides immediate feedback about their work.
- A World Wide Web communications system allows students to talk on-line with other students or with their professor and for the professor to communicate with the students.
- Web multimedia simulation experiments (Physlets) and other Web resources can be used for concept tests, homework, class demonstrations, and as pre-lab activities.

The authors have developed a variety of rich resources. As more users join the system and make contributions, the resource bank will increase. The resources have been and can be adapted in whole or in part to a variety of college learning systems. This Just-in-Time Teaching system has been developed by a talented group of physics educators. The entire system or parts of it will appeal to you and to your students.

* Larry Hedges, *American Journal of Physics* 46, 207–210 (1978).

Preface

Just-in-Time Teaching (JiTT) is a pedagogical strategy that succeeds through a fusion of high-tech and low-tech elements. On the high-tech side, we use the World Wide Web to deliver multimedia curricular materials and to manage electronic communications among faculty and students. On the low-tech side, we maintain a classroom environment that emphasizes personal teacher-student and student-student interactions. We combine these disparate elements in several ways, and the interplay produces an educational setting that students find engaging and instructive. The underlying method is to use feedback between the Web and the classroom to increase interactivity and allow rapid response to students' problems.

We have based most of the discussion in this book on physics because it is our primary subject. However, there is nothing in our underlying method that is specific to physics. Interactivity and responsiveness are applicable to any instructional setting, and student achievement and motivation are important in any subject. While any course can benefit from JiTT, it is easy to describe those courses that can benefit the most: *any course that students consider to be of secondary importance to their lives or their education*. Courses taken to satisfy requirements and courses taken by part-time students meet these criteria. We use physics examples because we are familiar and experienced with them. We hope that this does not put off instructors from other fields. We encourage others to adapt our ideas to their own subjects.

In this book, we have several goals. After a brief introduction (What Is JiTT?), we will argue that new methods are needed in the teaching of physics. If you already believe this to be true, you may feel free to skim Chapter 2 (Why Use JiTT?). In subsequent chapters, we will describe the underlying philosophy of JiTT: an effort to engage students by encouraging them to take control of the learning process and to become active learners. The core element of JiTT is the interactive lecture. Interactivity is established by using World Wide Web–based preparatory assignments that are due shortly before class. The instructors adjust and organize the classroom lessons based on the student responses "Just-in-Time." Thus, feedback between the classroom and the Web is established. Another example of our emphasis on interactivity is our use of multimedia-focused problems, in which the information necessary to solve the problem is not explicitly given in the problem statement. These problems are well suited to the JiTT paradigm and are based on media focused problems based on Physlets (java applets for Physics) developed by Wolfgang Christian at Davidson College.

In Section Two, we discuss implementation issues and techniques. Chapter 5 discusses the basics of organizing a JiTT course, and Chapters 6 and 7 detail two distinct versions of the interactive lecture.

Throughout this book, we will describe many classroom and Web-based activities that we use to enhance and extend the results achieved with the interactive lecture. While not "strictly JiTT," these elements are closely aligned with the goals of engagement and interactivity. We would like to stress that there is no unique JiTT method. We define JiTT broadly, as an effort to involve students by using feedback between out-of-class and in-class activities. We will provide detailed descriptions of two distinct implementations of JiTT; we encourage further innovation and invite collaboration.

Section Three of the book provides resources for instructors who are interested in implementing JITT in their classes. These include a sample set of WarmUps and Puzzles in Chapters 8 and 9, and a description of Physlets and sample Physlet Problems in Chapter 10. In addition, Chapter 11 provides basic information on scripting Physlets, and Chapter 12 by Larry Martin and Aron Titus provides background information on various Web-based communication tools used in the JiTT method. Section Three ends with a chapter of frequently asked questions, and responses to them, about the JiTT method.

Many of the WarmUps, Puzzles, and Physlet Problems provided in Section Three are included on Prentice Hall's Companion Website for Douglas Giancoli's algebra-based physics text, *Physics: Principles and Applications 5/e.* The site is located at

http://www.prenhall.com/giancoli.

These resources will also be available beginning in August, 1999 on Prentice Hall's Companion Web sites for *College Physics 4/e* by Jerry Wilson and Tony Buffa and *Physics for Scientists and Engineers 3/e*, also by Douglas Giancoli. The answers and teaching notes for the WarmUps and Puzzles in Chapters 8 and 9 and the sample Physlet Problems in Chapter 10 will be particularly helpful for professors who have adopted these texts and have their students use the corresponding Web sites. Because the questions and problems on the Prentice Hall sites will periodically be revised and updated, we are creating a correlation key between the specific WarmUps, Puzzles and Physlet Problems discussed in this book and those on the Prentice Hall Companion Websites. The key will be updated as needed, and will be available on the Web at

http://www.prenhall.com/allbooks/esm_0130850349.html.

There are a great many people and institutions that have contributed to our efforts, and we take great pleasure in acknowledging their support and their interest. At Indiana University Purdue University Indianapolis, we have received support, both financial and moral, from all levels of the university, including deans William

Plater, Erwin Boschmann, and David Stocum, and from the chairman of the Physics Department, Durgu Rao, who gave that most valuable commodity: time. We are also indebted to Kashy Valiyi and to the entire staff of the Center for Teaching and Learning. We have had many fruitful discussions with our faculty colleagues, particularly Steve Wassall and Fritz Kleinhans. In our recitation sections, we (and our students) have often relied on the efforts of Han Paik. We also benefited greatly from Dave Smith of the University of the Virgin Islands, who spent a sabbatical year teaching in our recitations, writing Web pages for our site, and discussing many of the ideas in this book during their formative days. We have also received invaluable assistance from many of our students, particularly our graduate student Stephen Schuh and our undergraduate mentors and computer specialists Robin Chisholm, Kelly Cook, Dawn Larson, Erin McGarrity, Debra Robertson, and Jerry Travelstead.

We would like to acknowledge the efforts of Mike Lee for helping to program and test Java programs at Davidson College, and we would like to thank Dr. Mario Belloni for many useful and stimulating discussions about the incorporation of Physlets with JiTT pedagogy and for testing Physlet-based WarmUps in his classes.

At the United States Air Force Academy (USAFA), we have received invaluable support and recognition from the former and current deans of the faculty, Ruben Cubero and David Wagie. We are very grateful to the Physics Department chair, James Head, for supporting and helping to establish JiTT within the department and for an early, insightful review of this manuscript. Fellow physics faculty member Tom Summers has provided moral support and creative advice and has done a great deal of work on JiTT assessment as well as adopting JiTT methods himself. Several other physics department faculty have also become particularly ardent JiTT supporters, and we are indebted to them as well: Rolf Enger, Joyce Collins, Greg Finney, Geoff McHarg, and Delores Knipp. We learned much about students and their learning from fruitful discussions with Terry McGrath. We are deeply indebted to work-study student Jennifer Robins and casual status lieutenants Donnie Starling and Knute Adcock for their considerable technical expertise and day-to-day HTML, CGI, and other support upon which we constantly relied. The USAFA preflights have been made possible on a large scale, across multiple departments and institutions, thanks to the development of a "preflight editor" by David Bell, Dana Kopf, and Mike Hawks. Randy Stiles and Carolyn Dull of the USAFA faculty Center for Educational Excellence have also provided key support for preflights at the institutional level. We are grateful, too, to Chris Wentworth at Doane College for helpful comments on this manuscript and for his efforts in developing a JiTT delivery system and database.

We would like to thank Jeanne Narum of Project Kaleidoscope for her enthusiastic support and for the many opportunities she has given us to speak publicly about our ideas and efforts. Many of our friends at AAPT have also given us recognition and support, especially Robert Hilborn, Nina Morley, and Paul Zitzewitz. Workshops have been an especially fruitful arena for the give-and-take of ideas with fellow faculty. We are particularly glad to have worked with Doyle Davis and

Francine Wald on many such occasions. The JiTT strategy could not have grown and matured without these opportunities and the exchange of ideas that they afforded.

Some people have been such frequent contributers of time and ideas that we have brought them in as the authors of Chapters 7 and 12 of this book. However, we would like to thank Rolf Enger, Larry Martin, and Aaron Titus again, both for their writing and for the many valuable ideas we have gained during our associations with each of them.

We would like to thank Laurie Gavrin for the many hours she spent applying her sharp eyes and precise grammar to the manuscript. Any mistakes that remain are likely the result of changes made since her last inspection.

All of us express our thanks to Alison Reeves of Prentice Hall, for encouraging us to write this book and convincing us that it could be bigger and better than any of us would have thought.

We also wish to express our sincerest thanks and apologies to those who suffered and supported us through many missed meals, late arrivals, and countless other inconveniences: our spouses, Miriam, Brian, Laurie, and Barbara.

The JiTT project has received support from an NSF grant (DUE-9554744) awarded to Doyle Davis for WebPhysics Workshops. JiTT is also partially supported by the current NSF WebPhysics grant to Davidson and IUPUI (DUE-9752365). One of us (WC) would like to acknowledge partial support during a sabbatical year from the U.S. Department of Energy and Duke University, and one of us (GN) would like to acknowledge support from the USAFA contracts office.

If I had to reduce all of educational psychology to just one principle, I would say this: The most important single factor influencing learning is what the learner already knows. Ascertain this and teach him accordingly.

D. P. Ausubel, in *Educational Psychology: A Cognitive View*

As you enter a classroom ask yourself this question: If there were no students in the room, could I do what I am planning to do? If your answer to the question is yes, don't do it.

Gen. Ruben Cubero, Dean of The Faculty, United States Air Force Academy

Increasingly, colleges and universities find themselves being asked how, when, in what settings, and through which methods and media their students learn best.

Policy Perspectives, 1998

The carpenter's attainment is, having tools which will cut well, to make small shrines, writing shelves, tables, paper lanterns, chopping boards and pot-lids. These are the specialties of the carpenter. Things are similar for the trooper. You ought to think deeply about this.

Miyamoto Musashi, in *The Book of Five Rings*

Section One: Strategy

Chapter 1: What Is Jitt?

Just-in-Time Teaching (JiTT) is a teaching and learning strategy comprised of two elements: classroom activities that promote active learning and World Wide Web resources that are used to enhance the classroom component. Many industries use Just-in-Time methods; they combine high-speed communications and rapid distribution systems to improve efficiency and flexibility. Our use of JiTT is analogous in many ways. We combine high-speed communications on the Web with our ability to rapidly adjust content; this makes our classroom activities more efficient and more closely tuned to our students' needs. The essential element is feedback between the Web-based and classroom activities.

We have built the JiTT system around Web-based preparatory assignments that are due a few hours before class. The students complete these assignments individually, at their own pace, and submit them electronically. In turn, we adjust and organize the classroom lessons in response to the student submissions "Just-in-Time." Thus, a feedback loop between the classroom and the Web is established. Each lecture is preceded and informed by an assignment on the Web. This cycle occurs several times each week, encouraging students to stay current and to do so by studying in several sessions that are short enough to avoid fatigue.

The JiTT Goals

We strive for both physics content mastery and the acquisition of more general skills. We also design our courses to provide experiences in teamwork and opportunities to practice written and oral communication. Our goal is to help the whole spectrum of students advance and learn, rather than targeting the average students or either extreme. The Just-in-Time Teaching strategy provides appropriate levels of support and feedback. JiTT provides remediation and encouragement to the weaker students while providing enrichment to the stronger students.

Students enrolled in a course that successfully implements JiTT will:

- ❑ Gain both problem-solving skills and conceptual understanding.
- ❑ Connect classroom physics to real-world phenomena and to their careers.
- ❑ Be in control of their own learning processes.
- ❑ Develop their
  - critical thinking ability.
  - ability to frame and solve problems.
  - teamwork and communication skills.

In addition to traditional homework assignments, students taking a JiTT-based course work in two interactive instructional environments. They work at their own pace in a virtual, Web-based setting that continually evolves with the progress of the class. They also collaborate with each other and instructors in a highly interactive classroom. Electronic communication among students and faculty provides a bridge between these two settings.

The JiTT strategy specifically targets obstacles facing many of today's students:

- ❑ Motivation to learn physics.
- ❑ Study habits and academic backgrounds.
- ❑ Confidence in their ability to succeed.
- ❑ Time constraints.

These goals and difficulties are addressed by combining high-tech (Web-based) and low-tech (classroom) elements, which we will discuss throughout this book. The feedback between these elements and between the people involved is the most fundamental asset of the JiTT method.

The JiTT Environment

We have been student-testing this strategy for five semesters and are encouraged by the results, both attitudinal and cognitive. For details, visit the JiTT Web site:

http://webphysics.iupui.edu/jitt.html.

In fact, working with the JiTT strategy has convinced us that the Web, combined with live teachers in the classroom, can humanize instruction for all students and make a real difference to the nontraditional student.

We have developed JiTT concurrently at three institutions: Indiana University Purdue University at Indianapolis (IUPUI), the United States Air Force Academy (USAFA) in Colorado Springs, and Davidson College. The JiTT strategy is effective despite numerous differences among the three institutions (we will elaborate in Chapter 2). This suggests that Just-in-Time Teaching is applicable in many other settings. The generality of the JiTT approach is also shown by our experiences with national JiTT workshops attended by faculty from a broad spectrum of institutions. For example, Daniel Kim-Shapiro, an assistant professor of physics at Wake Forest

University, a private four-year liberal arts institution, has successfully employed the JiTT strategy in his calculus-based introductory physics course taken by approximately 50 students, most of whom were pre-med majors. His students gave the use of the strategy an overall rating of 8 out of 10 on an end-of-course survey. It is interesting to note that in similar surveys at IUPUI, USAFA, and Davidson, our students also gave the JiTT strategy a score of 8 out of 10.

What JiTT Is Not

Despite our best efforts to explain what JiTT is, some readers may pick up false impressions. With this in mind, we would like to list a few things that JiTT is not:

JiTT is **not** a way to "process" more students.
JiTT is **not** "Just-in-Time Training."
JiTT is **not** distance learning.
JiTT is **not** computer-aided instruction.
JiTT is **not** designed from student evaluations.
JiTT is **not** market research.

We pay attention to our students' comments and suggestions. We agree with some but disagree with others. The JiTT strategy was not designed with student evaluations as the motivation for change; it was designed to address pedagogical issues.

Chapter 2: Why Use Jitt?

Many instructors in today's classrooms sense that students are not learning as well as they could. There is a mismatch between what we try to teach and what we find that students learn. This feeling comes to us as we talk to students in our offices and as we respond to their comments in class, and it is explicitly reinforced as we read their tests and quizzes. Education research results bear out this notion [Cope and Hannah, 1975; McKeachie et al., 1986; McDermott, 1991].

Our Students

There are many reasons why students are not learning as much or as well as we believe they should. Our purpose here is not to divide blame among students, faculty, and institutions. Rather, it is to improve the situation. We identify the primary underlying problem with those students who are not engaged with a course of study. For whatever reason, they are pursuing a passing grade rather than knowledge or ability. We proceed from this assumption and describe a method for reversing this attitude. The JiTT method motivates students to study the right things for the right reasons. It keeps them focused on knowledge, helps them to organize their efforts, and helps them anticipate success based on serious effort.

Thus far, we have tested this approach on as wide an audience as possible within the bounds of physics. The JiTT method has been jointly developed by faculty at three institutions: Indiana University-Purdue University at Indianapolis (IUPUI), the United States Air Force Academy (USAFA), and Davidson College. Our students and our institutions are diverse in background, goals, and the challenges they face.

IUPUI is an urban, public institution serving a largely nontraditional student body. These students have a wide range of backgrounds, skills, needs, interests, and objectives. Many of our students are returning to or first entering higher education after several years in the workforce. Most are the first in their families to pursue a higher education, and many have little confidence in their academic abilities. Some are on split schedules: in school in the morning, at work in the afternoon, and perhaps back in school at night. Many are supporting families.* As a result, most IUPUI students are well motivated and goal oriented, but they often carry enormous burdens on their time. Thus, they are suspicious of any course that is not clearly related to their career goals.

Retention is a major problem at IUPUI. Part-time students have a particularly hard time staying the course. They frequently give up early in the semester only to come back and try again the following semester. Some students question their ability and give up on their educational goals altogether. These reactions are not unique to IUPUI; high attrition rates are common among students returning to higher education nationwide [Cope and Hannah, 1975]. These retention problems can often be traced to time constraints and to limited feedback and support mechanisms outside the classroom. Some students who are under severe time constraints do not spend much time on physics outside of class; others spend quite a bit of time but find the time is not effective.

The USAFA differs from IUPUI in many ways. It is a moderate-sized rural military academy serving a student body composed of recent high school graduates who are bright and highly motivated. Retention of students is not a major problem. However, the cadets face many of the same challenges as IUPUI students. All cadets, regardless of their academic major, are required to pass one year of calculus-based physics. Those who are not physical science majors often question the relevance of this program to their military careers. Further, the cadets face severe time constraints due to an ever present combination of academic, military, and athletic duties. Although faculty are generally accessible during the day, time-constrained student schedules force much of their study into evening hours, when faculty support is not readily available. Regardless of their causes, the problems associated with severe student time constraints and limited after-class support have the same adverse effects on students.

Davidson College is a small, highly selective, private college of the liberal arts and sciences, with an enrollment of 1,600 students. Its students are also among the best in the nation. However, unlike IUPUI or the USAFA, Davidson does not feature a large engineering program. Introductory physics courses are dominated by students satisfying the college's core requirement for a laboratory science. These courses typically have an enrollment of 32 students split into two 16-student laboratory sections. Full-time faculty members teach both the lecture sections and the laboratories. Nevertheless, physics students at Davidson also benefit from and enjoy an interactive approach. The development of Physlets (flexible Java applets designed to illustrate physics concepts) and Physlet-based problems was pioneered at Davidson and will be described in detail later in this book. The computer-rich approach to JiTT pioneered at Davidson engages a wide variety of students with

* By contrast, a "traditional student" is the stereotypical college student who is 18–23 years old, has a solid high school education, and has large blocks of time available for study. Such a student has few responsibilities other than college and is financially secure. In science or engineering, this student would also most likely be male.

interesting and challenging material while attracting majors to the department early in their undergraduate career.

Our Goals

Science faculty are increasingly becoming aware of the challenges just described and of a need to expand the scope of our courses. Simply to present students with basic physics ideas and examples is no longer acceptable. To college and university graduates pursuing career objectives, the benefit of a college education is more than possession of knowledge in specific subject areas. A successful career in a technical workplace also necessitates the development of critical thinking, estimation skills, and the ability to deal with ill-defined problems. Equally important in today's society are cooperative work habits and communication skills. A recent American Institute of Physics survey [AIP Education and Employment Statistics Division, 1995] asked the employers of students who enter the workforce after obtaining an undergraduate physics degree what attributes and characteristics the employer most valued in their new hires. Ability to communicate effectively and ability to work well in a team topped the list rather than technical proficiency and competence, as might be expected.

Finding the Right Approach

In response to these pressures, many science faculty have re-examined their instructional objectives and have developed alternative pedagogical strategies. We have incorporated two prominent instructional innovations into the Just-in-Time Teaching strategy: active learning in the classroom [Sutherland and Bonwell, 1996] and the use of technological resources in support of teaching and learning.

Active Learning

It is now widely accepted that learning environments emphasizing student engagement tend to produce better prepared students than "traditional" courses, in which information is passed to students through lectures [Hake, 1998]. In the physics community, active-learner methods are frequently referred to as "interactive-engagement (IE)" methods, a term coined by Richard Hake. He defines interactive-engagement methods as

> ...those designed at least in part to promote conceptual understanding through interactive engagement of students in heads-on (always) and hands-on (usually) activities which yield immediate feedback through discussion with peers and/or instructors [Hake, 1998].

Studies of instructional methods suggest that improvements may occur if IE methods are undergirded by additional features, including "use of IE methods in all components of a course and tight integration of all those components, careful attention to motivational factors and the provision of grade incentives for taking IE activities seriously," augmentation of the teaching staff by undergraduate/graduate students, "apprenticeship education of instructors new to IE methods," "early recognition and positive intervention for potential low-gain students," and "more personal attention to students by means of human-mediated computer instruction in some areas" [Hake, 1998].

In a traditional lecture, the instructor presents new material by defining the terms, stating the principles, deriving the relationships, illustrating with examples, and showing demonstrations. Traditional lectures have a number of strengths [Cashin, 1985]: They can present large amounts of information to large audiences, they can model how professionals work, they allow the instructor maximum control of the learning experience, and they appeal to those who learn by listening. Lectures also have a number of weaknesses: They fail to provide instructors with feedback about the extent of student learning.* In lectures, it is difficult to intellectually engage the students. Hence, they have little choice but to learn passively, and therefore they retain less. Also, lectures are not well suited for teaching higher orders of thinking [Cashin, 1985]. The interactive lecture, which we define in Chapter 3, is designed to address several of these weaknesses. The students learn in an active environment, and the instructor gets constant feedback about the status of this learning.

Incorporating Technology

The education community has adopted various technologies as they have developed. The electronic-information age opened with the advent of the microprocessor-based personal computer in the late 1970s. The next phase came with the introduction of multimedia objects such as digital images, digital video, and sound. The third phase, the World Wide Web, has become the vehicle for rapid dissemination of multimedia learning resources and, for the first time, two-way communications. The Web technology provides for instantaneous interactivity, essentially eliminating the traditional space and time barriers. Web sites can serve as communication hubs for student-teacher, student-student, and teacher-teacher interactions. Web sites can also provide a virtual setting for the collaborative learning that has proved so effective in the classroom.

A student enrolled in a physics course that uses these instructional innovations works with many resources: the textbook, electronic course materials, the World Wide Web, the instructor, the teaching assistant, and other students. No technology can match the benefits a student derives from an expert human mentor who observes the learning activity, intervenes as needed, and thus makes the learning experience more productive and less frustrating. On the other hand, there are aspects of learning where technology can have an edge over a human instructor. A workstation has infinite patience, is nonintimidating, is nonjudgmental, and is available at all times for as long as the student wishes. This is particularly important to students who are facing major time constraints.

* "Thinking in terms of how much the student is learning as opposed to how much material has been presented is a fundamental and necessary shift in perspective" [Sutherland and Bonwell, 1996, p. 32].

Electronic mail provides a simple example. During off hours, a student can compose a far more detailed and coherent query using e-mail than could ever be possible in a telephone message; in return, a faculty member can offer an equally detailed electronic response. The entire "conversation" occurs without the need to match the participants' schedules and does not sacrifice either speed or accuracy.

Web-based interactive problems provide another example. Animation has tremendous potential, and it has been incorporated in Web documents for a number of years. However, it is often used ineffectually or for strictly cosmetic purposes. Merely including animation will not generally have a positive effect on student understanding [Titus, 1998]. For animation to be an effective educational tool, it too must foster an interactive-engagement approach. One promising approach is to embed general-purpose interactive programs in HTML documents and to use a Web-based scripting language to modify the behavior of these programs. We call these embeddable programs Physlets and are distributing them free to physics professors and instructors for use in their own classrooms [Christian, 1998]. Using Physlets and the feedback mechanisms inherent in a JiTT approach, students interact with the animations by making measurements and by writing about their observations. Using Physlets, we can pose new types of problems that would have been impossible to describe on the printed page. Requiring a student to select a lens to correct a nearsighted eye while observing the refraction of light rays produces a greater conceptual understanding of nearsightedness than solving the lens maker's equation.

Efficient use of the Web requires coordination. The JiTT approach combines communication technologies with in-class and out-of-class activities to optimize the values of each. Students interact with one another, with human instructors, and with electronic resources. JiTT provides feedback among these activities. In the following pages, we will describe the JiTT classroom and the JiTT Web component, with particular attention paid to the connections between the two.

JiTT in Context

Just-in-Time Teaching is an instructional strategy based on the active-learner approach to instruction [Sutherland et al., 1996]. Student teams take charge of their own learning. Faculty experts and student mentors facilitate the learning process supported by electronic communication technology. Thus, JiTT is another implementation of the new ideas that are fueling higher education transformation and revitalization efforts [Pew Higher Education Roundtable, 1998]. In the parlance of physics teaching, JiTT is an interactive-engagement (IE) approach to instruction, as described by Richard Hake [Hake, 1998]. In these methods, expert instructors engage the students in the learning process; the students maintain primary control of their learning while interacting with instructors who help them to stay on task and advance. To summarize Richard Hake, such a system of instruction cannot succeed unless a minimal set of conditions is met: IE methods must be employed in all components of a course, and all of these components must be tightly integrated. Careful attention must be paid to motivational factors, and grade incentives must be present to encourage students to take IE activities seriously. Undergraduate and/or graduate student mentors must augment teaching staff. Personal attention must be given to students by means of human-mediated computer instruction.

All of the well-known and widely acclaimed recent initiatives in physics education centrally employ IE methods. Among these are Peer Instruction [Mazur, 1996], Workshop Physics [Laws, 1991], and Tutorial-based instruction [McDermott, 1991].

In Peer Instruction, a portion of the classroom time is spent with students trying to answer and understand the physics content in carefully crafted multiple-choice questions. The instructor presents an unfamiliar question to the class and asks the students to take a short amount of time to try to answer the question individually. After the deliberation period, the students make their choices known to the instructor by some means such as raising cards or inputting their choices into an electronic system. The whole class is informed of the overall distribution of selections. Next, the students are given a few minutes to discuss their choices and their reasoning with their neighbors. Students then submit another answer. Based on the final distribution and the details of the question, the instructor then gives a brief lecture that brings out the finer points of the question, focuses on stumbling points, considers extensions of the original question, etc. One can easily imagine using the JiTT feedback before class to select which Peer Instruction questions to incorporate into that lesson, and, in fact, some JiTT adopters have already found the combination of JiTT and Peer Instruction to be particularly effective. One can even construct JiTT questions with a certain set of Peer Instruction questions in mind and then pick from that set based on the JiTT student feedback obtained. Eric Mazur has adopted some of the JiTT techniques in his introductory courses at Harvard and reports that Peer Instruction works well in concert with JiTT. He assigns WarmUp questions before each class and responds to student needs and questions via a combination of e-mail and in-class attention. [Mazur, 1998 private communication]. As another example of the importance of communication, Mazur notes that the student feedback he has received from using a JiTT approach to reading quizzes has helped inform refinements to his upcoming calculus-based introductory physics text.

Workshop Physics is a "hands-on, minds-on" activity-based student-centered approach to introductory physics pioneered by Priscilla Laws at Dickinson College. The Workshop Physics curriculum is research-based and is intended to help students develop their conceptual understanding of physics, be able to relate that understanding to mathematical representations, achieve wider scientific literacy, develop their data-taking and analysis skills, and improve their motivation to learn science [Laws, 1997]. In Workshop Physics, students participate in a variety of research activities, such as making qualitative observations of actual physical phenomena, many of which involve a kinesthetic component; gathering and analyzing data from digitized videos and/or computer-based data-acquisition laboratory tools; modeling data using spreadsheets; problem-solving; guided equation derivations; and discussions with classmates and instructors. In the Workshop Physics approach, students first observe or experience a physical phenomenon and then are guided to construct a model that explains their observations. By including a variety of activities, the curriculum appeals to a broad spectrum of student learning styles. The curriculum is centrally based upon the notion that "students learn physics when students are doing physics" [Laws, 1997].

Introductory physics students are rarely "empty vessels" who have no preconceived notions or prior experience with a large part of the typical introductory physics curriculum; using JiTT techniques to probe their understanding prior to the classroom experience can provide a valuable segue into the classroom activities. Indeed, Professor Laws has begun incorporating some JiTT methods into her introductory physics courses at Dickinson (private communication, 1998).

The Tutorials approach to introductory physics is based on an extensive and detailed body of research into student preconceptions or misconceptions about physics. Student responses to written pretests and questions and/or activities during student interviews with physics education researchers have brought forth a wealth of information about what students do and do not understand, both before and after traditional instruction. Based on this research, a series of student activities called tutorials has been produced by Lillian McDermott and the Physics Education Group at the University of Washington [McDermott and Shaffer, 1998]. The tutorials are intended to supplement the lectures and textbooks used in "traditional" courses with the goal of improving student learning in these courses. At the University of Washington, traditional recitations (which generally featured a physics graduate student solving textbook problems on the blackboard) have been replaced by tutorial sessions. During the tutorials, students in small groups work through paper-and-pencil worksheets. Carefully coached teaching assistants and instructors rove the room, observing the groups and intervening as necessary. Proper preparation of the TAs and instructors is a key component of the tutorial-based system, and, in fact. there are weekly seminars for those in teaching roles. The tutorials are designed to actively engage and intellectually involve students in their own learning process to help them develop a functional understanding of physics.

At the USAFA, an interesting experimental introductory course incorporating both McDermott-type tutorials and JiTT was under way for the first time during the fall 1998 semester (Maj. Greg Finney, course director, Physics 215Z, Fall 1998, private communication). Since the tutorials place heavy emphasis on conceptual understanding, the JiTT WarmUp assignments used for nontutorial class meetings tend to emphasize problem-solving skills and techniques. In preparation for a tutorial class session, however, the WarmUp generally includes questions that connect naturally to the tutorial activities. In this course, students are better prepared for the tutorial experience because they have thought about the material, and instructors gain valuable insight prior to the class.

There are many other active-learning innovations in physics education, such as the Socratic dialogue methods introduced by Richard Hake [Hake, 1987; Hake, 1991] the Active Learning Problem Sets (ALPS) introduced by Alan Van Heuvelen [Van Heuvelen, 1991], and context-rich Problem Based Learning [Heller, 1992].

These all rely on students being in control of their own learning, guided by a human instructor who questions the student and provides support as needed.

The JiTT implementations described in this book contain elements of all of the approaches described above, tailored to our particular students. We have added as a crucial feature a rapid-cycle feedback engine, which has only very recently been made possible by the ubiquity of World Wide Web communications. The addition of feedback loops can pay big dividends in any IE approach without disturbing the unique features that make it valuable to the students for whom it was created. Appropriately adapted, components of JiTT as described in this book may well enhance the effectiveness of any active-learner student-centered pedagogical technique, in any field or subject area. So far, we have seen indications that JiTT can be combined with many other active-learning techniques. In each case, the result is an environment that is even more supportive of student learning. Members of the JiTT team have given workshops to physics and nonphysics faculty, a good number of whom have adopted JiTT techniques, sometimes in conjunction with other IE approaches.

Since the primary emphasis of the JiTT strategy is on the preparation of the student and the teacher for the classroom encounter and for subsequent follow-up, there is no reason why it should be more applicable in one discipline than another. We hope that JiTT will not only thrive across the spectrum but that it will lead to a dialogue between teachers and students across the disciplines and thus improve learning for all.

Chapter 3: The JiTT Classroom

The JiTT method is most closely associated with the interactive lecture, which we will describe below. However, the underlying philosophy remains engagement and interactivity. In the following sections, we will describe two kinds of classroom activities. The first activity is the interactive lecture, and the second is a collaborative recitation session as we practice it at IUPUI. The recitation is not "strict" JiTT: There is no electronic assignment that precedes and informs the lesson. However, this session is highly interactive and works well in connection with the JiTT lecture. Most of the ideas in this book should be taken as guidelines rather than as strict solutions. Just as we remind our students that problems may be solved by many methods, we remind our readers that the JiTT approach to teaching has many possible variations. JiTT can be adapted to existing courses in a variety of ways. For instance, Physlet-based JiTT problems were added as pre-labs at Davidson College because they fit into Davidson's three-hour laboratory time slot without major curricular revision. Since faculty were already comfortable managing an interactive learning environment, the JiTT approach to pre-Labs was quickly adopted by the entire Davidson physics department.

The Interactive Lecture

The interactive lecture session, described extensively in the "JiTT Implementations" part of the book, is intimately linked to the electronic preparatory assignments the students work on outside of class. This session, where the instructor takes the opportunity to talk, is built around student answers to questions known as "WarmUp Exercises" at IUPUI or as "Preflights" at the USAFA. These are short Web-based assignments that are due just hours before class; students submit their answers electronically. We design these assignments to encourage the students to assemble any prior knowledge they have about the upcoming topic. The typical student needs to preview the material in the textbook in order to complete the assignment. The instructor collects the students' electronic submissions, reads them, and presents excerpts from them during class, weaving them into the lesson as appropriate. Thus, the students take part in a guided discussion that begins with their own preliminary understanding of the material. We do not simply "go over" the student responses in an isolated section before or during a lecture; rather, we frame our lecture and classroom activities in terms of an analysis of various student responses.

It is important to understand that our use of the Web is not intended to reduce the demands on instructors' time nor is it intended to increase the number of students that can be handled. If anything, the role of the instructor and the involved commitment to individual student needs become even more important.

The primary goals of the WarmUp exercise and interactive lecture system are to:

- ❑ Encourage students to prepare for class regularly.
- ❑ Help teachers identify students' difficulties in time to adjust lesson plans.
- ❑ Help students develop a stronger "need to know."
- ❑ Establish an interactive environment in the "lecture" classroom.

It may appear that this method puts an unbearable stress on a new faculty member who is struggling with research goals in addition to lesson preparation and classroom management, but an instructor committed to an interactive classroom would be fielding many questions and facilitating classroom discussions anyway. Acquiring information about the students' levels of understanding and students' questions before class allows the faculty member time to reflect and prepare rather than being surprised by "out of the blue" questions in class. A discussion about managing an interactive classroom and detailed examples of how we use the WarmUps/Preflights are provided in Section Two (Implementation), and a sample set of WarmUp/Preflight questions is provided in Section Three (Resources). A complete selection of WarmUps, Physlet problems, and other resources may be found on the companion Web site to Douglas C. Giancoli's text *Physics: Principles with Applications*. This site may be accessed at

http://www.prenhall.com/giancoli.

Two JiTT Lecture Implementations

In chapters 6 and 7, we present detailed examples of two implementations of the interactive lectures, along with specific examples of WarmUp and Preflight assignments. We will discuss our motivations in developing the materials and operating the lectures according to these two methods. We show actual samples, actual student responses, and present a detailed discussion of the classroom activities that we constructed based on these materials.

The first implementation assigns relatively more weight to developing the connection between physics theory and real-world phenomena, and to the development of technical communication and teamwork skills. We label this approach the "Mindful" interactive lecture, as it draws on Ellen Langer's notion of mindfulness (discussed below). The second implementation assigns more weight to developing conceptual understanding, insight, and the operational skills needed to solve problems at the introductory level. We label this approach the "Insightful" interactive lecture to emphasize the value of this strategy in helping both teachers and students explicitly gain insight about textbook weaknesses, student misunderstandings, and gaps in student facility with prerequisite material.

Of course, these weighting systems are actually dynamic; the emphasis may shift during the semester. When progress slows because many students have trouble with a basic skill, the emphasis on connecting theory to the real world is reduced in favor of focusing on the tools, e.g., vector calculus in parts of E&M.

The Collaborative Recitation

The sessions start with a 10–15-minute review of the homework problems. The emphasis is placed on systematic problem-solving techniques and a few "physics highlights," the fundamental ideas common to each of the problems. In contrast to what occurs in recitation sections at many institutions, detailed solutions to the problems are not presented.

The remainder of the class is devoted to collaborative problem solving: students are given a set of problems that they have not seen before. In the recitation sessions at IUPUI, 60–70 students meet in a common room and work in groups of two to four students. They do their work on full-sized whiteboards that have been installed around the periphery of the room. The student teams first discuss the approach they will take. Then they work toward a solution to the problem. During this time, faculty members and student mentors circulate among the groups, giving guidance only as needed and observing the students as they work. At IUPUI, two faculty members staff these sessions, along with several undergraduate student mentors (usually recent students who are good communicators and have a desire to teach). A third professor or a graduate student is also sometimes present. At the USAFA, all classes are small enough for one faculty member to facilitate (15–24 students).

This format has several advantages in teaching introductory classes. Because students must tackle unfamiliar problems, faculty may address problem-solving skills that beginning students often lack. Because students work in groups, they must attempt explanations of their work to their peers. This is often the point at which a student realizes that a particular method or idea is more complex than previously thought. Further, a faculty member may observe that a student or group has a basic misconception and step in to provide one-on-one or small-group instruction.

This system establishes close, informal interactions between faculty and students, and these interactions have benefits that extend to other aspects of the course. The faculty and teaching assistants get to know the students personally, observing their strengths and weaknesses. Armed with this knowledge, they can better help the students optimize their study time. Students who would normally hesitate to come to a faculty member's office have far less compunction after only a few classes. The familiarity spills over into the lecture as well; the willingness of students to ask and respond to questions during lecture is dramatically improved. This is particularly beneficial in establishing the interactivity so valued in the JiTT lecture. In addition, the student-student interaction is invaluable in maintaining student morale. In the virtually anonymous setting of a commuter campus, it is all too easy for a student to despair and drop the class under the (often false) impression that they are hopelessly behind everyone else. Even at the USAFA, where all the students are in residence, the student-student communication helps them see that they are all engaged together in the business of learning. More details about managing the collaborative recitation session appear in Chapter 5 (JiTT Basics).

We would like to stress that this organization of classroom activities is only one of many possibilities. JiTT is a general method, adaptable to many possible implementations. As an example, at IUPUI, the formal lecture-recitation-lab meeting structure is still in place; the lecture and recitation occur on alternating days. At the USAFA, the classes are small and meet in rooms designed for multiple purposes; there, the various activities are typically blended together in each class meeting. At Davidson College, PreLab assignments similar to WarmUps are used to improve student learning in the laboratory sections.

Chapter 4: The JITT Web Component

Through systematic use of the emerging network technologies, it is possible to break down the space and time barriers and connect to students outside the classroom at any time. The Web component of the course gives students more control over their learning processes, provides information connecting the course material to real-world phenomena, enhances the students' sense of belonging to a learning community, and provides remedial materials when necessary. Furthermore, the Web-based assignments are carefully constructed and paced so as to motivate students and help them manage their study time.

The JiTT Web material facilitates an effective blend of human and technological resources. The Web assignments encourage the student to prepare for the classroom activity, where the instructors will provide the intellectual mentoring that only a human being can deliver. The Web material also encourages the student to think beyond the limits of the course material. Providing a map to the constantly growing wealth of information available on the Web is an important function of the JiTT Web site. Electronic submissions encourage clarity and conciseness in writing, an important skill for the workplace. We require that the exposition be in plain English. It must include supporting arguments referring to the physics and mathematics but should do so without the use of equations. This is how an engineer or a scientist would communicate with the nontechnical members of a project team. The instructor shares (anonymous) elements of the submissions with the class. This establishes a feedback loop that is beneficial to students and instructors alike. Discreet in-class critique of these submissions improves both the writing and the supporting physics and mathematics evidenced in subsequent submissions.

The JiTT Web materials fall into several broad categories: electronic preparatory assignments and enrichment extra-credit materials (either of which may include Physlet problems), on-line homework, and information and communication. Each of the Web assignments has different educational goals. However, the Web assignments also work together in several respects. By scattering the credit over several different pages, we encourage students to work on physics frequently and in short sessions, a far more effective study habit than a single weekly (or less frequent) "cram session" [Britton, 1991]. The Web component also provides several new communication channels. Since students respond to these assignments electronically, they are in frequent one-on-one contact with their instructors. Requests for clarification of a particular topic often accompany their submissions. Furthermore, the course Web site includes an electronic bulletin board, allowing students to contact one another outside of class. These communication tools also serve to further strengthen the students' sense of a learning community. This aspect is particularly important at IUPUI, where most students are commuters.

Many of the materials that we have developed for use in the JiTT system may now be used by students and faculty using Giancoli, *Physics: Principles with Applications 5/e* through Prentice Hall's Companion Web site. The site is accessible at

http://www.prenhall.com/giancoli.

On that site, the essay and estimation sections of the WarmUp exercises and the Puzzle questions (described below) are all grouped together under the heading "warm-ups," and Physlet problems may be found under the heading "Physlets" for each chapter. The "What Is Physics Good For?" essay and associated questions may be found under the heading "applications."

Preparatory Assignments

There are two primary categories of electronic preparatory assignment: the WarmUp/Preflight and the Puzzle. These assignments provide an introduction and a conclusion to a given topic. The preparatory Web assignments share a common set of characteristics:

- The questions asked are motivated by a clear set of learning objectives.
- The Web assignments introduce students to the technical terms.
- The students confront their previously held notions.
- The questions are extendible.

Helping students recognize that their "everyday" understanding of words and ideas may be incorrect or incomplete is a crucial part of this process. Furthermore, reminding them to reconsider and reflect on actual experiences they may have had, and asking them to confront their preconceived ideas about the way nature works, helps students to construct new and deeper reasoning and understanding [Fuller, 1982]. In fact, this idea is central to the Learning Cycle, a teaching strategy developed by physicist Robert Karplus [Karplus, 1977]. The JiTT Web component helps students consider their everyday understanding and experiences, realize that their current mental constructs are incomplete, and come prepared to deal with them in the classroom, where they are guided by the instructor. As an example, consider "moment of inertia." First, the student must realize that there is nothing momentary about it and that the object may have zero inertia. Only then can the instructor introduce the physics, preferably using a scenario where it is obvious to the student that both "how much stuff is there" and "how the stuff is distributed" will affect the outcome of a process. To attain a working knowledge of the subject, students must be able to deal with the jargon (specific physics terms) and with everyday words or phrases that can be used with nonexperts. It is important that the questions be extendible so that during classroom discussion previous material can be revisited. Thus, students gain new insight, and future topics can be anticipated, as illustrated in the example below.

The WarmUp

The WarmUp (Preflight at the USAFA) affords the students some ownership of the "lecture" session. These exercises are the heart of the Just-in-Time instruction system. Before each class, the instructor has a few hours to read the student responses and adjust the lesson in response to the students' demonstrated knowledge. An excerpt from a WarmUp exercise is shown in Fig. 4.1.

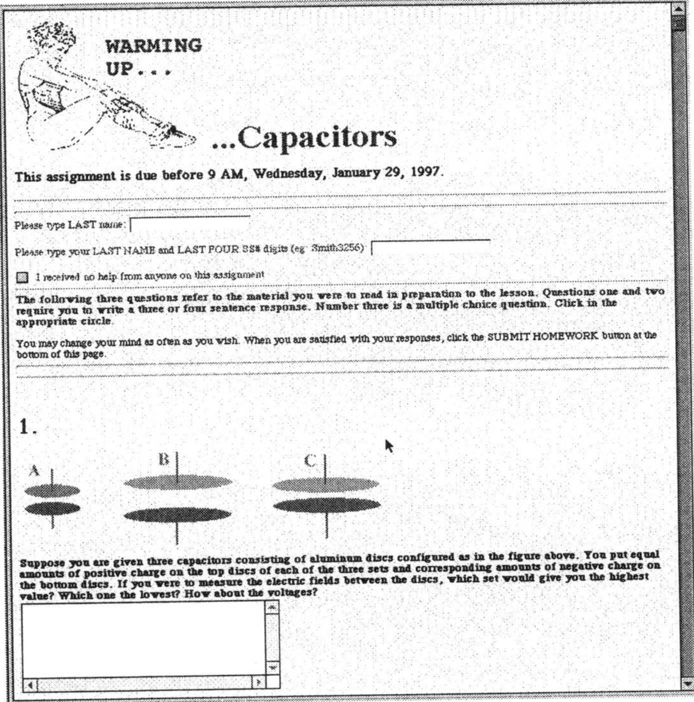

Figure 4.1 The beginning of a WarmUp exercise from the introductory Electricity and Magnetism course at IUPUI.

We have found that conceptual questions are particularly valuable. Most instructors (ourselves included) easily forget the level of understanding that students bring to their first contact with various concepts. Consider the words "force," "power," and "energy." In colloquial English, these are synonymous, but in our physics classes we give these words special meanings and insist that the quantities not be equated. Students, pressed for time, often miss this fact on a first reading of the book. A WarmUp question asking the students to write about these concepts will motivate many students to reread a few sections of the text. Those who base their answers on prior ideas also benefit. When the question is discussed in class, they are more inclined to listen when the phrase "Many of you wrote something like this..." appears prominently.

The students' answers (presented in anonymous, edited form) are used as talking points for the instructor, while the issue is still fresh in the students' minds. By using this method, the instructor can "individualize" the lecture. There is no content sacrifice since many of the responses are predictable to an experienced instructor. The important point is that students sitting in the classroom recognize their own wording, both correct and incorrect, and thus become engaged as part of the feedback loop. It is quite common for the classroom discussion to continue via e-mail between the instructor and particular students. Clearly, this interaction and engagement does have a cost. Contrary to the common perception that the use of technology in teaching reduces instructor time, the JiTT use of technology actually presents a greater demand on instructor time. Paradoxically, technology used this way encourages a more personal and intimate bond between instructors and students. It is clear from course evaluation responses that students feel part of a team working on a common project. Everyone has a stake in the final outcome and, as a result, the overall performance improves. Class attendance and participation goes up, the attrition rate goes down, and, we believe, real learning increases.

Details about the kinds of questions we include in WarmUps, the motivations behind our choices of questions, and how we actually structure the follow-up classroom activities are provided in Section Two (Implementation).

The Puzzle

Each week we assign one "Puzzle," which is due at the same time as the first WarmUp: a few hours before the first lecture of the week. Typically, the puzzle is a single question that may be somewhat vague and usually involves several concepts. As the name implies, the puzzle is a physics scenario with an extra twist that requires the student to see beyond the end-of-chapter formulas. This is often accomplished by further challenging commonly held misconceptions. The Hestenes' Force Concept Inventory questions [Halloun and Hestenes, 1985] and Eric Mazur's peer instruction questions [Mazur, 1996] have a similar character. These questions may appear trivial to a physicist, and many can be answered without calculations. On the other hand, they usually present a significant challenge to the beginning student.

In the classroom, we use the puzzles much as we do the WarmUps. Based on excerpts from the students' solutions, we lead a classroom discussion of the question and of possible variations and extensions. We usually use this exercise to close a topic and to integrate it with the rest of the course material. We think of the

WarmUp and Puzzle activities loosely as bookends for a given topic, beginning with the WarmUp and ending with the Puzzle.

Physlet Problems

Physlets are small, scriptable Java applets capable of displaying physics content. Because Physlets are embedded into HTML documents and because they are scriptable, they can easily be used in concept tests, homework, prelabs, and in-class demonstrations. Physlets are easy to use since they are based on standard non-proprietary Internet technologies. Many examples of the use of Physlets may be found on Prentice Hall's Giancoli Companion Web site. For example, Prentice Hall's website for Giancoli, *Physics 5/e* has approximately 8-10 problems per chapter that are based on the Physlets. (Note that the problems on this and other Prentice Hall Web sites are copyrighted by Prentice Hall, and may not be used without permission; however, instructors may download the underlying Physlets from the Davidson College Web site for non-commercial use in their own classes.) The URL for the Davidson site is

http://webphysics.davidson.edu.

A technology should not be used if the pedagogy it produces is not sound. We would like to distinguish between media-enhanced problems, where multimedia is used to present what is described in the text, and media-focused problems, where the student uses the multimedia elements in the course of solving the problem. Multimedia-focused problems are fundamentally different from traditional physics problems, and Physlets are ideally suited for these types of problems. For brevity, we will refer to these as Physlet problems. Consider this example: a traditional projectile problem states the initial velocity and launch angle and asks the student to find the speed at the maximum height. This problem can be media-enhanced by embedding an animation in the text, but this adds little to the value of the problem. Alternatively, this same problem could be a Physlet problem. In this case, no numbers are given. Instead, students observe the motion, apply appropriate physics concepts, and make measurements of the parameters they deem important within the Physlet. Only then can they "solve the problem." Such an approach is remarkably different from typical novice strategies where students attempt to mathematically analyze a problem before qualitatively describing it (an approach we as teachers often call "plug-and-chug" and characterize by the lack of conceptual thought during the problem-solving process). Requiring students to consider the problem qualitatively has a positive influence on students' problem-solving skills and conceptual understanding.

Consider problem 11 from Chapter 5 of the Giancoli Companion Web site, shown as a screenshot in Fig. 4.2. The student is asked to find the "planet" that does not obey Kepler's laws. How does a student solve this problem? The student must observe the motion and recognize that the ratio of the square of the orbital period to the cube of the orbital radius is required for a number of planets. Numerous orbit parameters must be measured, and it is unlikely that the first measurements will yield anything unexpected. In fact, the orbit that "looks" most unphysi-

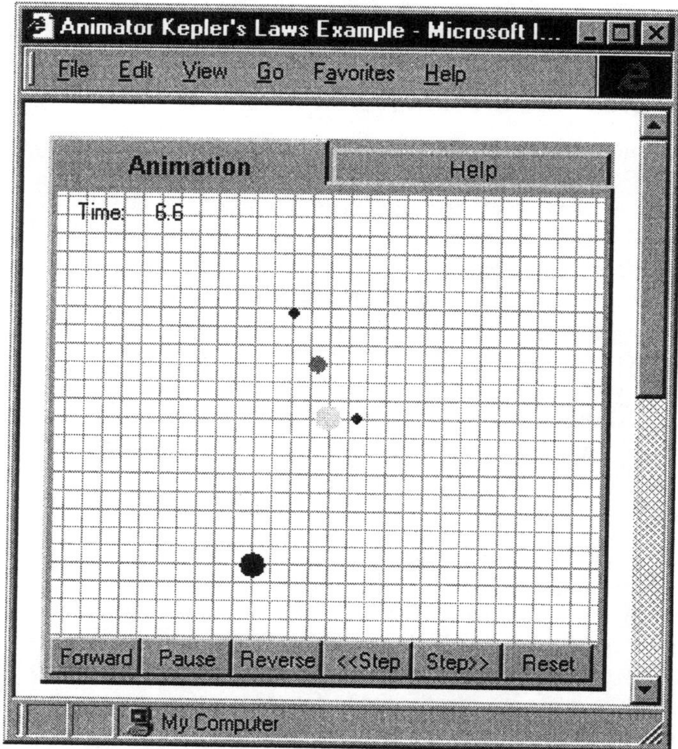

Figure 4.2 A Physlet that shows a simulated solar system in which one planet has a nonphysical orbit.

cal is the outer orbit because, the planet moves very slowly. The unphysical innermost planet zips along at too slow a rate. This visual representation of abstract formulas is surprising to students.

How is the Physlet problem different from a similar traditional problem? In a traditional problem (see Giancoli, *Physics: Principles with Applications,* 5th Ed., p. 142), the student is given an orbital parameter for a satellite and is given another satellite for comparison, thereby suggesting the path to the solution. In comparison, the multimedia-focused problem requires observation and conceptual reasoning before quantitative analysis. Observation of the Physlet also reinforces the idea that the outer planets move very slowly in comparison to the inner planets.

We believe that multimedia-focused problems are closer to the real world than traditional textbook problems. Upon first encountering multimedia-focused problems, many people comment, "They are like virtual laboratories." As in an experiment, students must determine what they need to know to solve the problem before attempting a solution. Likewise, the answer depends on the method of solution and experimental error. Instructors can use Physlets to ask questions similar to those encountered in actual laboratory situations. Physlet problems are ideally suited for use in JiTT, as prelabs, for example. More details on the use of JiTT prelabs can be found in Chapter 5 under the heading "The Small-College Perspective."

Enrichment Materials

We noted earlier that the collaborative recitation is not a "strict" JiTT activity (as it does not rely on a preceding Web question). Similarly, the "What Is Physics Good For?" Web page does not usually have a formal in-class component. However, like the recitation, this Web page serves to motivate and engage students. Indeed, although it is rarely discussed during class, it often dominates discussion *before* class, as students arrive and settle in.

The "Good For" page brings a weekly essay on some topic related, but not central, to the course material. The essays fall into several categories. Among these are important physics-related news events (e.g., the launch of the Mars probe); everyday phenomena (lightning, rainbows); historical perspective on the ideas of physics (energy, atomic theory); technology (electronics, civil engineering); and devices in everyday life (batteries, refrigerators, car engines). Students enjoy these and constantly suggest new topics and categories.

These essays differ from the "application" essays found in physics textbooks in several important ways. They contain numerous links to external Web sites where a wealth of related material is available. Also, the essays always end with extra credit assignments that induce the students to follow the Web links and thus relate the course material to the real world. We refer to these assignments as "extra credit" for a specific reason. When we began using the Web, we initiated the distribution of T-shirts for the first correct answer to each puzzle. While several highly motivated students vied for this prize, most students ignored the Web site altogether. After informally interviewing several students, we concluded that students would not visit the site unless there was a direct academic incentive. We did not wish to reduce the amount of traditional homework, but students will resent extra "required" work if it is imposed with no concomitant increase in the number of credit hours earned. Offering "extra credit" answers both of these concerns, while simultaneously improving student morale.

Student comments from end-of-course surveys indicate that these essays are extremely well received (their overall rating was 8.7 out of 10). Even if they didn't have time to try to earn any of the extra credit offered through these pages, many of the students made time to read the essays because they reinforced the real-world connection and illustrated the relevance of the course material to the students' lives.

Information and Communication

The course Web site also contains several pages that serve as communication channels. These range from simply providing electronic versions of traditional handouts to more sophisticated interactive tools. "This Week in Physics" is a newsletter-style page that keeps students informed of happenings in the class and often calls attention to physics-related news events. Quite frequently there is a midweek edition of the page, e.g., after a major test or an important news event. This serves to maintain a steady "drumbeat" that paces the students, reminding them of assignments, sources of help, etc. The information pages contain the electronic editions of the course policies, syllabus, and calendar.

The communication pages provide students with several opportunities to communicate among themselves and with faculty. There is an anonymous electronic "suggestion box" which faculty monitor regularly, as well as faculty phone and e-mail information. There is also an electronic course bulletin board, which students may use to communicate among themselves. This is particularly useful during the early weeks of the semester when the students do not know one another well. At these times, the bulletin board is often used to set up study sessions and to exchange e-mail addresses. Other pages offer archives of previous materials and a "credit check" that allows students to determine which Web assignments they have received credit for and to gain a sense of how they are doing with respect to their classmates. The home page of one of our course sites is shown in Fig. 4.3.

In addition to the other communication tools, many of the assignment pages carry a "general comment" box. The WarmUp page comments box has proved especially useful since it provides a place where a student can express concerns, frustrations, or positive feelings immediately after they have completed an assignment. These "free form" comments are an invaluable indicator of the general climate in the classroom and provide yet another vehicle for improved communication between faculty and students.

On-line Homework

Homework is an import part of the JiTT teaching strategy. Including both exercises and problems, it plays a prominent role in almost all curricula. Exercises are defined as assignments where the answer is already known and the emphasis is on learning a process. Typical exercises might include resolving vectors into components or solving simultaneous equations. Few teachers would want to cover motion in two dimensions or Kirchhoff's rules if students do not have the prerequisite skills. Problems, on the other hand, are assignments where the answer is not known ahead of time. They seek to develop new concepts or to use and combine existing concepts in new ways. WarmUps and Puzzles presented previously are examples of problems. We have found an automated on-line grading system to be very effective in assessing student performance on both types of homework and in keeping faculty informed of student progress. However, it is crucial that faculty also read a representative selection of student responses. The JiTT method works best when faculty members understand their students' progress qualitatively as well as quantitatively.

Figure 4.3 The home page of Physics 152, the introductory mechanics course at IUPUI. The navigation bar at the left provides easy access to "This Week," Warm-Ups, Puzzles, etc.

Automation of routine tasks should be accomplished whenever possible and, since JiTT requires additional work on the part of faculty for class preparation and curriculum design, routine grading is an obvious target. We have tested an automatic on-line homework system, WebAssign, for routine grading at Davidson College. Although our initial interest was to use technology to increase the time available for student-faculty interaction, it soon became clear that there are also pedagogic advantages to this approach that make it particularly compatible with JiTT. There has always been a fine line between students working together—a process we encourage—and plagiarism. Even tutors in departmental homework-help centers have a hard time knowing where to draw the line. Creating random values for each problem's initial conditions for each student is an effective way to encourage group learning while still requiring individual effort. Students can talk about problems and even share techniques, but each answer is unique. Copying numbers into someone else's formulas is still, of course, not learning. A student who engages in this practice would likely plagiarize work under any system and will likely fail the in-class test.

We have set our system to allow unlimited submissions before the assignment due date. Students are not penalized for getting a problem wrong and are told which answers are correct. They do not, however, know the correct answer, as with back-of-the-book solutions, and thus cannot work toward a known number. A scan of the answer log shows that many students do a quick run through an assignment in order to determine which problems they understand and which problems require further study. The remaining problems often have four or five incorrect submissions before they are successfully completed. The automated system gives students immediate feedback; this allows students to self-evaluate and to concentrate their effort on those problems where they have the most difficulty. Assignments are usually due one class period, i.e., two days, after we have covered the relevant material in class. A faculty member can quickly pull up a summary organized either by student or by problem. Students stay on track, and the next lecture can, if needed, be adjusted if a particular problem presented widespread difficulty. Finally, a Web-based system can deliver media-focused problems that are unavailable in a traditional text.

Automated homework grading has not decreased student traffic during office hours, but students do seem to come by earlier in the homework cycle. For example, in the past at Davidson, homework would be collected at the beginning of class, graded, and returned within two class periods. Even though homework keys were posted immediately, students would typically wait until an assignment was returned before asking for help on an incorrect problem. In addition, it was impossible to collect and grade all assignments since over 200 problems are given in a typical semester. (Davidson physics faculty also grade their own tests and student laboratory notebooks.) Students now seek help before an assignment is due since they already know that they are having problems. Homework grades have, in fact, improved over the previous system. We consider this to be a good thing because we believe that this accurately reflects the added effort being devoted to understanding the material.

There are numerous homework systems available on the Web, and many of them are suitable for JiTT. We have chosen to use WebAssign at Davidson College. WebAssign was designed by Larry Martin and is currently marketed by North Carolina State University as a homework-grading service for departments that do not want to manage their own servers. A noncommercial version of WebAssign is currently being written as part of the WebPhysics project. It will be downloadable from the Davidson College Web server:

http://webphysics.davidson.edu.

Additional details of Web-based technologies such as CGIs, Java, and JavaScript can be found in Section Three (Resources).

The Links Between JiTT Components and Instructional Objectives

The Web and classroom components each serve their own purposes, but the combined ensemble works to address the course goals, instructional objectives, and particular challenges outlined above. The table below summarizes the degree to which the primary components address or target the key course objectives and challenges.

Relative Importance of Various Jitt Methods to Our Educational Goals

Goal	Web components				Classroom activities	
	WarmUp	*Puzzle*	*Good For*	*This Week*	*Recitation*	*Interactive lecture*
Motivation	moderate	**high**	**high**	moderate	moderate	moderate
Real-world connection	moderate	moderate	**high**	moderate		moderate
Critical thinking	**high**	**high**	**high**	moderate	**high**	**high**
Estimation	**high**	moderate	moderate			moderate
Ill-defined problems	**high**	**high**	moderate			**high**
Teamwork					**high**	
Communi-cation	**high**	**high**	**high**	moderate	**high**	moderate
Time management	**high**			moderate	moderate	**high**

Section Two: Implementation

Chapter 5: JiTT Basics

As described in Section One of this book, Just-in-Time Teaching consists of a carefully orchestrated blend of learning activities. Students perform some of these activities at their own pace, on their own time. The preparatory exercises are delivered via interactive World Wide Web documents. The setting for the second set of activities is the classroom, where instructors and students, working in teams, construct the knowledge the students take away from the course. The classroom activities are intimately tied to the preparatory Web activities. Students do the lion's share of the work; the instructor's job is to author (or select from a database) the appropriate set of Web documents, consider the student responses, and guide the subsequent classroom activities.

The key is to use Web communication technology to prepare the students and the instructors for the events in the classroom. This gives the students an active role in the process. The day-by-day logistics will depend on the schedule of classes, but any instructor with an interest in using an interactive paradigm can adapt JiTT to his or her course structure.

JiTT Organization

As an example of scheduling a course using JiTT, consider the introductory mechanics course for science and engineering majors at IUPUI. The catalog lists the course organization as two lectures, two recitations, and one two-hour lab per week. The JiTT version of the IUPUI course is illustrated in the diagram.

The class meets every day, and there is an assignment due every day. Most of the assignments are submitted electronically from Web pages that are always available. The exception is paper homework that is due, for credit, twice a week just before the recitation. The recitation is described in Chapter 3. Fig. 5.1 illustrates the flow of the various assignments into the classroom environment.

The daily aspect of the work has two important advantages. It helps students manage their time effectively. If homework is due only once each week, many (perhaps most) students will do it during one or two marathon sessions. Spreading the work out is also helpful because it allows students to feel that they are "keeping up" with the course. At any given time, they can look at the next assignment and

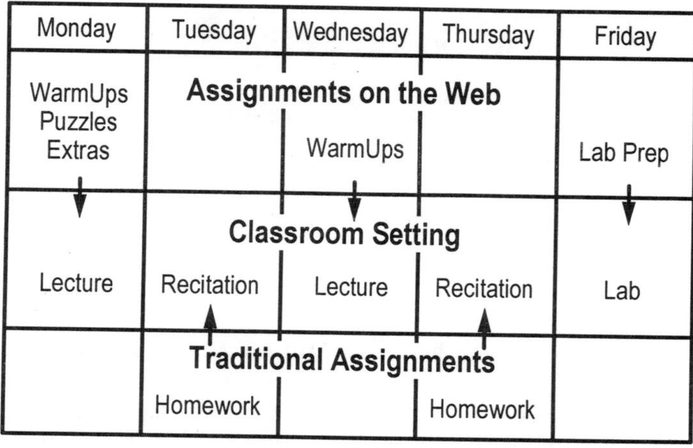

Monday
Tuesday
Wednesday
Thursday
Friday
WarmUps
Puzzles
Extras
Assignments on the Web
WarmUps
Lab Prep
Classroom Setting
Lecture
Recitation
Lecture
Recitation
Lab
Traditional Assignments
Homework
Homework

Figure 5.1 The flow of student assignments into the classroom setting in Physics 152 at IUPUI.

find it to be manageable. Of course, students do find the pace difficult, most miss assignments on occasion, and some initially voice complaints. However, most students benefit from this strategy. One student put it in this way:

> I think the warmups are a good idea because without them most students would not even look at the material before class.... I think I'm a better student because of warmups!... I think [the Web-based assignments] are additional teaching formats and I think that it has kept me focused on physics because I always had something due every day.

The Web assignments are an integral part of the course. Through their responses, the students largely determine the way we present the physics in the classroom. The Web materials prepare students for the subsequent in-class activities. On the WarmUps, students get credit for effort, not for content. The course grade is still based mainly on traditional in-class tests. However, there is enough credit (about 20% at IUPUI) available on the Web to motivate most students; over 80% do the assigned Web work.

Note that the elements of the traditional lecture-recitation-lab format are still present. The new element is the student involvement in the classroom activities. Students' need-to-know and interest are piqued because, by the time they come to class, they have already grappled with the material.

Managing the JiTT Classroom

Initially, the least well-defined and therefore most daunting aspect of JiTT is classroom management. In the JiTT faculty workshops we have led, the bulk of participant questions deal with practical matters: how to involve as many students as possible, how to lead effective discussions, how much time to spend on the Web-based homework responses, etc. Walking into class armed with insider knowledge about what the students do and do not understand can be a tremendous advantage. But efficient and effective use of the classroom time does not necessarily follow just because the instructor has insight into the students' understanding. Since the key to the JiTT strategy is the feedback loop between the Web-based homework and the interactive classroom session, careful management of the classroom is essential for continuing successful JiTT implementation.

David Hestenes, the principal investigator on NSF teacher-enhancement grants for high school physics teachers for nearly a decade, recently commented [Hestenes, 1998] on what makes a good teacher:

> Managing the quality of classroom discourse is the single most important factor in teaching with interactive-engagement methods. This factor accounts for wide differences in class FCI score among teachers using the same curriculum materials and purportedly the same teaching methods. Effective discourse management requires careful planning and preparation as well as skill and experience.

Richard Hake further comments that there seem to be no "magic bullets" with regard to interactive-engagement methods of teaching, but based on his extensive survey [Hake, 1998] he suggests that

> ... improvements may occur through, e.g., (a) use of IE methods in all components of a course and tight integration of all those components; (b) careful attention to motivational factors and the provision of grade incentives for taking IE activities seriously; (c) administration of exams in which a substantial number of the questions probe the degree of conceptual understanding induced by the IE methods; (d) inexpensive augmentation of the teaching/coaching staff by undergraduate and postdoctoral students; (e) apprenticeship education of instructors new to IE methods; (f) early recognition and positive intervention for potential low-gain students; (g) explicit focus on the goals and methods of science (including an emphasis on operational definitions); (h) more personal attention to students by means of human-mediated computer instruction in some areas; (i) new types of courses; (j) advances in physics-education research and cognitive science.

The interactive-engagement nature of JiTT suggests that these instructional strategies or pursuits are factors that are highly relevant to a successful JiTT implementation. Indeed, we believe that the JiTT strategy explicitly considers and addresses each of the factors articulated by Hake.

Perhaps the single most important factor governing the effectiveness of the classroom time is the overall climate in the classroom. JiTT works best when the instructor promotes a friendly, relaxed atmosphere in which the students feel they are part of a large team (including the faculty) working together towards a common goal. If the students believe that their preclass efforts help in this endeavor, they have additional motivation to complete the Web assignments, and they feel more ownership of the classroom time as a result.

How does one promote such a climate? Experience in a variety of JiTT settings suggests taking a substantial amount of time right at the beginning of the semester to explicitly address the motivations and the day-to-day procedures associated with the course. This brings the students into the process from the outset, highlights their important role in the learning process, and underscores the vital nature of their participation.

Managing the Interactive Lecture

Let's assume that at this point you're a faculty member who has been able to construct and offer good JiTT questions to your students and that they have been able to submit their responses to you well before class. You have chosen some sample responses that set the stage for a discussion of the points you wish to make. In many cases, you will have made transparencies, handouts, or electronic versions of the representative submissions to help your students focus on the specific wordings and explanations. You may also have printed out your student submissions and annotated the printouts. Furthermore, you may have calculated some rough statistics on numbers of students who responded in a given way (correct, incorrect with a particular mistake, etc.) so that you can be sure to let students know where given responses stand relative to the class. You have probably also sketched out an outline for your classroom time. Armed with all this, you walk into your classroom. What will actually transpire?

The JiTT instructor must strive to involve as many people as possible in the classroom discussions and/or activities to best capitalize on the student ownership factor. For example, before discussing a particular WarmUp question, the instructor can repeat or show the actual question so that even those students who were unable to see and answer the question before class are able to participate in the classroom activities. Mentioning the percentage or number of students who responded to a particular question at all and then mentioning the approximate percentages of responses that fall into particular categories based on the nature of the answers can also help to draw many of the class members into the discussion. For example, referring to a particular anonymous response that is being displayed for the students, "All but three of you answered this question, and of those of you who did, just over half of you said something very similar to this response...." In such a case, the three nonparticipants would probably feel a bit left out and would typically (in our experience) react to this feeling by making an extra effort to respond to the next Web assignment. Also, the whole classroom of people will be more likely to pay attention to the ensuing discussion once they hear that a large fraction of the students answered in a particular way. Was that correct? If not, why not? Such a "mainstream answer" must be worth analyzing.

The best way to encourage nonparticipants is to convince them, through the classroom discussions and activities, that they are missing out on something by not participating. The important point here is that the discussion of the WarmUps (and the Puzzles) *is* the lesson. The development of the physics theory is intertwined with the WarmUp questions and answers. These serve as springboards for the introduction of the lesson's topics. The temptation to keep the WarmUp discussion separate from the "lecture" must be avoided if this method is to work. This holds to an even greater extent for Puzzles. The Puzzles typically involve a mix of concepts the students have just finished studying; a thorough discussion of the puzzle is, in fact, a review lesson.

The students must be keenly aware that the discussions and activities in the JiTT classroom stem from, and are centered on, their actual responses. If this is not the case, the crucial feedback loop between the Web and the classroom components is broken. On the other hand, they must also be aware that there is an underlying sense of organization and structure to the classroom time so that the class time does not seem like a rambling, unorganized session of reading and critiquing student responses. Striking the balance between these factors can be challenging. It takes deliberate and intentional organization and considerable thinking and planning before class. As he or she is reading the student responses before class, the JiTT instructor must know what key points must somehow be brought in via those student responses and comments. The classroom time must then unfold so that these points are clearly and logically raised and discussed, using the student responses as the springboard.

We have found that a good way to do this is to be clear and explicit with the students about the purpose(s) of the classroom discussions and activities. For example, "As you probably recall, today's first WarmUp question deals with a merry-go-round. From our discussion of this today, we'll end up defining the angular kinematic quantities, working through the equations described in your textbook reading, and figuring out how and when to use those equations." Frequently ask (and discuss) the questions, "Why are we doing this?" and "How does this relate to today's topic?" Also ask, "Have we seen something like this before?" Questions of this kind help students make connections to earlier material.

The JiTT instructor works through the lesson material by asking questions and soliciting student answers to those questions in class. The first of these questions is almost always about a particular sample student (Web) response to the first WarmUp question.

What kinds of student responses make good discussion starters? Common responses are very useful because of the ownership issue mentioned above and because they permit the instructor to deal with the perspective of many of the students in the room. Also, responses that provide good details and explanations are generally more useful than those that are vague and unsubstantiated. An instructor might make an exception to this by deliberately highlighting a student response that demonstrates essentially correct understanding but is not well enough explained. This provides an opportunity to discretely critique the writing style, draw out class comments about the response, and suggest opportunities for improvement in the quality of the answer.

Typically, it is not useful to pick an "off the wall" response to highlight in class unless it has some unique appeal, such as humor, or a totally innovative interpretation of a question. We have found it is usually best to react to such responses in a one-on-one fashion, either by exchanging e-mail with the student, or engaging him/her in a before- or after-class discussion of the response. Try, when possible, not to let such responses go unanswered or unaddressed.

As points are brought out through the classroom discussion, the JiTT instructor provides structure and organization to those points. We believe that careful faculty boardwork is important, as it helps the students formulate the framework for the knowledge they are constructing. Key points, symbols, and equations can be transcribed in logical locations and in a premeditated order on the board. Once again, careful planning before class time is the key to ensuring that this actually happens in a way that helps the students.

It is important that the classroom time be highly interactive, with as much student participation as possible. Even if it is based on the student Web submissions, a straight lecture that could occur with or without the students actually being present is not in the spirit of continually striving to engage the students in the learning process.

To facilitate interactivity, the JiTT instructor may ask open questions directed at anyone who chooses to answer. Open questions are fine, but it's important to allow plenty of time for someone to volunteer an answer! Allowing sufficient "wait time" is often difficult for an instructor, particularly if he/she is new to interactive teaching methods. Generally, it takes a few long pauses (at least five seconds) after posing questions early in the semester. Soon the class realizes how important their participation is and that the instructor will wait until there is some kind of response—even a question for clarification—from a student. Once this happens, the wait time required shortens considerably. Of course, instructors can also choose to explicitly discuss the fact that they will wait for a response, rather than just answering their own questions. This also has the effect of shortening the wait time.

The JiTT instructor may also ask questions directed to a particular student. Some faculty very successfully use directed questions to help the entire class stay focused on the current discussion.

Another approach that encourages student engagement is to ask a particular question or series of questions and then ask the students to discuss the questions in groups of two or three. After one to two minutes, the instructor asks for reports of the student discussions. This technique is used to great advantage by Eric Mazur [Mazur, 1997]. This technique has the advantage of a high fraction of the students having actually thought about the questions and formulated answers, rather than just a few students who are "on the spot" because they volunteered to answer questions or were called upon. The student in-class discussions are also fertile ground for generating new discussion points and student-offered extensions.

Where possible, entertain student-proposed extensions to questions posed. This encourages student participation in class and encourages students to make connections between the current material and their everyday experience. It also helps them to connect with previous topics. If a student raises a point that is related to the topic at hand but not absolutely central to it, guiding comments like "Remember where we've come from" or "Remember our starting assumptions" can help to frame the subsequent discussion and allow the students to make connections. Determining how far afield to let the classroom discussions wander and how many extensions to offer really requires a dynamic decision based on the class pace and comfort level with the material.

What about questions that are too far afield from the topic at hand? Typically, one might first praise the student for asking a good question, albeit it one that is not quite appropriate for that particular class session. If the question will naturally be answered in a subsequent class, the instructor can mention this fact to the student and suggest a one-on-one session later if he/she wants to talk about it sooner.

As important as student involvement is, success is not ensured just because an "interactive lecture" is achieved. One pitfall to avoid is having only one or two students dominate the discussion while the rest of the class sits by. Both nonparticipants and dominators are dangerous to the JiTT classroom environment. Often, an off-line face-to-face conversation with a particularly eager or assertive student is all that is required to correct a classroom imbalance in participation. It is well worth having such a conversation as soon as the situation is detected! The entire class will benefit as a result.

Demonstrations and Other Activities

Often, even a well-conducted and lively discussion of a specific point is not sufficient for student mastery of that point. Classroom activities can and should be more than discussions. A well-chosen demonstration or student activity can go a long way toward solidifying new ideas and notions about concepts. For example, consider the WarmUp question about stepping onto and off a merry-go-round. "What happens if you step off a rotating merry-go-round platform? Is the speed of the platform affected? How?"

After the classroom discussion of that question, a good demonstration might be a student sitting on a platform that is free to rotate, holding weights in his/her hands with arms outstretched. What will happen if the platform is spun up while the student holds his/her arms out and then he/she drops the weights? How is this similar to the WarmUp question scenario? How is this different? What do you predict will happen? Why? In this case, the demonstration is used as a follow-up activity, and it represents a (slight) extension to the original discussion. Other possibilities for demonstrations and student activities include:

- things done before discussion, to lead into the discussion
- things done that represent exact reinforcement of the actual question discussed
- video clips that reinforce the discussions

Managing the Collaborative Recitation

Unlike the interactive lecture, where the instructor leads the session, the recitation session consists mostly of self-paced student teamwork. The facilitators for this session are a mix of faculty members, graduate students, and undergraduate student mentors. Students appear to appreciate the diversity of the facilitating group. We agree with the Hake observation [Hake, 1998]:

> I have found top-notch undergraduate physics majors, after suitable apprenticeships, to be among the best IE instructors, evidently because their minds are closer to those of the students and they have only recently struggled to understand introductory physics concepts themselves. Thus they can better appreciate the nature and magnitude of the intellectual hurdles and ways to overcome them.... As future professionals, the undergraduate, graduate, and post-doctoral student instructors all provide the opportunity to seed interactive-engagement methods into science education at all levels.

To prepare for the session, students are assigned a set of regular homework problems from the book (not less than two, not more than five). These are handed in and are graded (at least spot graded) by the next class meeting. At the beginning of class, one of the instructors reviews the homework problems. This brief introductory session lasts not more than 20 minutes. The students are already familiar with the problems, so we use these examples to formally teach problem-solving strategies. After the short session on the homework, student teams work on problems they have not seen before.

Forming the Groups

At the first class meeting, students are instructed to assemble in groups of three or less, standing at the white boards which cover the walls of the lecture room. These initial groupings are tentative as, with some exceptions, students typically do not know one another. The groups form spontaneously. We typically do not use any of the recommended schemes (e.g., Heller, 1992) for forming these groups. We value the morale of the class highly and do not wish to jeopardize it by seeming overly intrusive, even if some benefit may be gained in other areas.

About three-quarters of the groups keep their members for the duration of the semester. The rest regroup within two or three weeks. There are always a few "floaters" who are slower to find a group with which they are comfortable. Sometimes one or two students prefer to work alone. By midsemester these students usually also find a group to which they can adjust.

Our class surveys indicate that the vast majority of students prefer collaborative problem-solving sessions to a lecture on problem solving. In every class, however, there are a few individuals who do not take well to collaboration. Frequently these students can get a very good grade on their own. In a typical recitation class, they hand in the homework, sit through the introductory part when the instructor discusses the homework problems, and then leave. At IUPUI we do not require them to stay for the teamwork session.

The Teamwork Problems

After the review of the homework problems, the student teams are assigned two or three more difficult problems. The teamwork problems are challenging but within the grasp of students working together. They always deal with the same physics as the homework problems but may integrate some earlier material. They are always multistep problems. All the groups work on the same problem set.

The teams are instructed to first develop a strategy to tackle the problem. As the groups work, the faculty and student mentors observe the groups. Ideally, they intervene only if asked to do so or if they note evidence of a serious misconception. Students who have made minor errors will generally benefit from being allowed to discover this on their own. We encourage the students to finish the problems and come to closure after class if necessary. This gives the students an opportunity to extend their group work beyond the class setting. We have identified several key issues:

- ❑ Is the problem laid out properly?
- ❑ Is there a diagram if appropriate?
- ❑ Has the relevant physics been identified correctly?
- ❑ Has the question been stated in physics terms (e.g., "police car has caught up" becomes $x_1 = x_2$) ?
- ❑ Has the problem been divided into sub-problems when appropriate?
- ❑ Are students entering numerical data prematurely?
- ❑ Are the team members sharing ideas or working the problem individually with little or no interaction?
- ❑ Does the team have a dominating member who simply does all the work or acts as a lecturer?
- ❑ Does the team have a weak member who adopts a secondary role (e.g., "scribe" or "calculator operator")?
- ❑ Is the group stuck? Are they trying random methods of attack?
- ❑ Once they solve the problem, are all the members comfortable with the solution?

We have also identified several pitfalls that facilitators should learn to avoid.

- ❑ Do not give "minilectures."
- ❑ Do not intervene needlessly. If they can find their own mistakes, let them.
- ❑ Work with a group's method even if there is a more efficient method.
- ❑ Guide their steps; don't carry them.
- ❑ Modify their work; don't substitute your own.
- ❑ Make positive statements about their efforts (e.g., "This is a good diagram. Let's rotate it to …").

Typically, students can finish two to three multistep problems in 40 minutes. At IUPUI we always assign three to keep the faster groups busy and to encourage the team to work on the third one outside of class. This keeps the team members working together, gives them a sense of accomplishment, and generates office visits and e-mail messages. In short, it provides another opportunity to extend the classroom.

The time spent with the problem-solving teams in this setting is valuable to the instructor. Face-to-face experience with teams supplements the insight gained through the student responses to the WarmUps and Puzzles and the subsequent interactive lecture discussions. The recitation experiences often suggest new Warm-Up questions and puzzles.

The Small-College Perspective

JiTT is easily adapted to different types of academic institutions. Small liberal arts colleges have historically valued student mentoring, small class size, faculty-student contact, and a sense of community and shared purpose. Much of the feedback mechanism that makes JiTT so valuable exists without a need for new technology; student lives are already centered on campus activities rather than on family or work, and faculty have daily contact with students. Students drop by the office for a chat, come to departmental picnics, bring their parents by for a visit on Parents Weekend, and socialize with faculty at campus events. Extracurricular problem sessions are easy to organize on a residential campus with a traditional 18- to 21-year old student body. Whereas JiTT can help establish and enhance the sense of community at a large urban university (such as IUPUI), at small colleges, such as Davidson College or the USAFA, we have emphasized its role as a vehicle for curriculum reform through technical innovation. Classroom management, motivation, and uneven performance are common problems even at the best liberal arts colleges, and the JiTT paradigm is effective in using technology to deal with them.

We have tested, adapted, and adopted the strategies of the interactive lecture and the collaborative recitation at Davidson College. However, not every MWF 50-minute lecture has a "classic" JiTT assignment that is due a few hours before class. There is little point in—and students have told us that they resent—having too many activities due in one week. They are, after all, taking three to four other classes, and the flood of information can overwhelm both students and faculty. We have found three or four on-line activities per week to be reasonable and attainable. In addition to pre-lecture JiTT exercises, we use weekly pre-laboratory assignments, occasional post-laboratory assignments, and end-of-chapter Web-based homework to close the feedback cycle. All these activities are blended into a JiTT strategy that fits well into a traditional curriculum.

JiTT in the Laboratory

Physics pre-laboratories have been used at Davidson College since before the advent of the World Wide Web. Their intent was—and is—to ensure that students come to laboratory prepared. In the past, they included the derivation of equations needed during the laboratory or the calculation of results using sample data. These pre-laboratories were the entry tickets to the laboratory. They accounted for 10% of the laboratory grade, and students could not begin the laboratory until the pre-laboratories had been handed in to the instructor. But they were only marginally effective in preparing students for the ensuing experimental work. The instructor graded them after the laboratory had been completed. This provided little guidance as to the student's understanding of the material during the laboratory. Furthermore, they were analytical and could just as well have been assigned as an end-of-chapter problem. Numerical values were presented in the handout; observation and analysis were not required.

Pre-laboratories have now been redesigned as JiTT exercises. They are presented to students on the Web and must be answered at least two hours before the beginning of laboratory. This allows the instructor to scan student responses and to begin the laboratory with a short discussion based upon student understanding. Alternatively, the instructor can visit a lab bench and work one-on-one with a student; in this way technology, allows us to personalize instruction even in a small class.

Web-based prelabs have another important benefit. We can now ask media-focused questions using Physlets. Media-focused questions require that students observe an animation and make measurements using that animation. Numerical values are rarely given. Sometimes these measurements mimic the experiment to be performed, but not always. For example, the conservation of momentum prelab presents students with an animated collision showing two identical balls. The collision is scripted so that it does not conserve momentum but does conserve energy, and the prelab asks if the collision obeys the laws of physics. The choice of what to measure is left to the student. Although such questions are rich in possibilities, they require only a minimal user interface. The Physlet displays time in the upper left-hand corner of the animation, and the student can pause the animation at any time using an on-screen button. The students can also measure positions of objects using a click/drag of the mouse.

This type of problem does not replace the laboratory. Rather, by presenting erroneous physics, it elicits the idea that observation is required in order to determine if theory is correct. The animation looks believable; only a trained observer would notice that something might be amiss. Multimedia-focused problems such as this force students to think about the purpose of the laboratory and encourage them not to assume forgone conclusions. In this case, for instance, most experimenters find that some momentum is lost due to interaction with the air track.

Chapter 6: The Mindful Lecture

In this chapter, we define an approach we call a "mindful" lecture, borrowing the term "mindful" from the work of Ellen Langer [Langer, 1997]. Langer offers the following definition of "mindfulness" in her book *The Power of Mindful Learning*:

> A mindful approach to any activity has three characteristics: the continuous creation of new categories; openness to new information; and an implicit awareness of more than one perspective.... Mindlessness, in contrast, is characterized by entrapment in old categories, automatic behavior that precludes attending to new signals, and by action that operates from a single perspective.

Note that a few of the goals of mindfulness were present in the traditional passive-listener mode of teaching physics. The content of early chapters would be seen in new light as students study later chapters. The material studied in an undergraduate course would begin to make sense in graduate school. In a Just-in-Time setting, all the aspects of mindful learning are explicitly emphasized. In this setting, instructors are also mindful learners. They "attend to new signals" coming from the students and adjust their actions accordingly. Students feel that the course is a team effort.

A mindful lecture is a team activity in which students and instructors work together to develop the physics content. By the time the students come to class, most of them have done substantial preparatory work. The instructor has posed questions on the Web that provide a context for the new material. Students will approach these Web assignments in various ways. Some will read the book and then attempt to answer the questions. Some will look at the questions and then go through the book trying to find useful information. Others will try to answer the questions without ever reading the book. The approach we encourage is to read the questions, read the book, and try to make meaningful connections. The students submit their work electronically before class, the instructor reads their submissions, and the whole team starts the classroom session on the same page.

The Mindful Lecture WarmUp

The WarmUp exercise gets the students grappling with the new physics and serves as a springboard before class. A WarmUp assignment consists of three parts: a short essay question, an estimation question, and a multiple-choice question. These questions are on material that has not yet been discussed in class. Sometimes the three questions deal with three different aspects of the same concept. Sometimes, though, they are related to three different objectives for a given lesson.

As mentioned above, we expect the students to do the assigned textbook reading and complete the WarmUp before class. The WarmUp questions encourage students to consider real-world scenarios and analyze them in terms of the physics content they are learning. The student chooses among the possible methods of understanding the scenario. This approach is similar to what Ellen Langer terms "sideways learning" [Langer, 1997]. This is in distinction to both the top-down method, which is essentially "lecturing to instruct students," and the bottom-up method, "repeated practice of the new activity in a systematic way."

> Sideways learning aims at maintaining a mindful state.... The concept of mindfulness revolves around certain psychological states that are really different versions of the same thing: (1) openness to novelty; (2) alertness to distinction; (3) sensitivity to different contexts; (4) implicit, if not explicit, awareness of multiple perspectives; and (5) orientation in the present. Each leads to the others and back to itself. Learning a subject or skill with an openness to novelty and actively noticing differences, contexts, and perspectives—sideways learning—makes us receptive to changes in an ongoing situation. In such a state of mind, basic skills and information guide our behavior in the present, rather than run it like a computer program.

To encourage maintaining a mindful state, the WarmUps explicitly draw attention to details that students often overlook. For example, consider the following WarmUp question:*

> The power loss in a resistor with current I and voltage V is given by $P = IV$. If the resistor obeys Ohm's law, $V = IR$, the power is also given by $P = I^2R$ or by $P = V^2/R$. Does that mean that P is proportional to R, inversely proportional to R, neither, or both? Explain.

Students who answer this question by casually (mindlessly) quoting formulas will become confused by what the algebra seems to imply. A mindful approach to this question requires the students to understand Ohm's law as a proportionality between V and I in a resistor and to apply this understanding to the expression for power loss in a resistor. Drawing explicit attention to relationships like this and asking students to contemplate them as well as to apply them helps the students internalize details in a more meaningful way. This fosters less superficial and longer lasting learning.

The purpose of the WarmUps is to "prime the pump" for learning; they are not assessment quizzes. Students earn credit based on their level of effort, not the correctness of their answers. Most students put forth conscientious efforts to answer

* This question and other dc electronics questions may be found at http://www.prenhall.com/giancoli by selecting Chapter 18 warm-ups.

the questions to the best of their ability. They quickly learn that their honest attempts to answer the WarmUp questions are rewarded by classroom discussions that relate to what they answered. In the discussion, we may praise student answers either for mastery of the content or for good communication style. However, we do not disparage incorrect answers. Rather, we may describe such answers as "incomplete," "sensible, but containing a misconception," etc. This may sound "soft," but recall that these are intended to be preliminary exercises, covering material that has not yet been introduced in class. We also do not present our own "correct answers" either in class or in print. Students often object to this, but we do not want archives of these available to future students.

The following few sections describe the three kinds of WarmUp questions we use in the mindful lecture implementation of JiTT.

The Essay Question

The goals of the essay question and the subsequent classroom activity are for students to learn the underlying physics theory and to learn to apply it to real-world problems. We pose the essay questions in everyday English and construct them around a real-world scenario. Engineers and scientists must develop analytical and communication skills. They should be able to convert real-world problems into models, apply physical theories, and express their results in either technical or nontechnical language. Ordinary citizens, too, should develop "scientific awareness—an understanding of what the scientific enterprise is about ..." [Devlin, 1998]. A good scientist or engineer knows the relevant physics, can identify the important parameters, can make meaningful judgments about which parameters to ignore, and knows how to break up the problem into subproblems without failing to solve the whole. Constructing physics knowledge in the context of real-world problems rather than bare physics theory ensures that these issues don't get lost. For example, a winning game of golf depends more on the Bernoulli equation than on simple projectile motion kinematics. The essay question also requires students to develop and practice their technical communication skills. We encourage them to verbalize the relationships between different quantities, not to quote formulas mathematically.

Student responses require both conceptual understanding and analytical ability. There are several facets to this process. Students must become aware of the physical quantities embedded in the text; a skier becomes a position, or an electron becomes a charge and a mass. Next, they must restate the question, using the technical terms to convert the question to a well-defined physics problem. After solving the problem, they must translate their results back to everyday language. Students are expected to use complete English sentences in their responses.

The students accomplish these tasks with varying degrees of proficiency and completeness. Since students working on their own complete only a subset of these tasks, the submissions serve as preparation for classroom discussion of this question. During the discussion, the instructor and the students will develop the physics content. A classroom session conducted in this manner covers the same material as would be covered in a traditional lecture setting, but the quality of the outcome is superior. Because student submissions shape the development, the students pay more attention, gain deeper knowledge, and retain it longer.

The Estimation Question

The second question in the WarmUp is an estimation question in which key information is missing. The goal of the estimation question and the subsequent classroom activity is for students to practice framing and resolving ill-defined problems, to enhance their critical-thinking skills, and to develop an intuitive feel for the abstract physical relationships they are learning. Many of our students begin the course with a poor understanding of the term "estimate." To some students, estimate is synonymous with "guess"; to others, it is equivalent to "calculate." Still others believe that an estimate must not involve the use of equations.

We construct the estimation questions with some data purposely omitted or given implicitly. In order to answer the question, the students must estimate the value of some quantity, but they must first decide what quantity or quantities they need to estimate. The missing information falls into two categories. Students can find data in the first category by looking it up in reference material like the textbook. Information of this kind, e.g., the radius of the Earth, has the same value for all students. Other data depend on choices made by the student, e.g., the weight of a car. Reasonable values for such data fall in a range. Often, the follow-up classroom discussion explicitly addresses how to determine reasonable values for missing data of this type.

Answers to the estimation questions can be either qualitative or quantitative. For example, "Can you and your family lift your car?" is a qualitative question, even though a quantitative estimate is required, whereas "Estimate the hang time of a basketball player" requires a quantitative response. In their responses to the estimation question, we ask students to state their assumptions, to state the values they supply, and to indicate what equations they used (but not reproduce the equations symbolically). They should give their numerical answer and comment on its reasonableness. If they cannot completely answer the question, they are expected to explain what they do and do not understand about the problem so that their responses provide some indication of their thought process and trouble spots they encountered.

The in-class follow-up to the estimation question often leads into a discussion about how to estimate a quantity whose magnitude isn't known. Occasionally extending this discussion to other estimation problems, often called Fermi problems, increases motivation and arouses curiosity in some students.* A note of caution is appropriate here. While some students, typically engineering and science majors, find these sorts of problems engaging and intriguing, many of the students, called

*A famous Fermi problem is "Estimate the number of barbers in Chicago." For more such problems, see http://physics.umd.edu/rgroups/ripe/perg/fermi.html.

the "second-tier students" by Sheila Tobias [Tobias, 1990], do not. The next section provides a specific example of an estimation question and a detailed description of how we discuss and extend this type of question in the classroom.

The Multiple-Choice Question

The third question of the WarmUp is a multiple-choice question. In many instances, multiple-choice questions are the most concise way to induce a fruitful discussion. Unlike the essay or estimate formats, a multiple-choice question can point out the richness of a physical situation and drive students to consider many possibilities. Furthermore, because it is possible to include choices that "sound good" but play to students' misconceptions, these questions can be used to lay traps that students would resent on a test but will remember in the classroom.

These questions are excellent discussion starters. The richness of the situation can be explored by considering what is wrong with each of the incorrect choices. Further discussions arise from considering the conditions under which each choice could be correct. Choices that are based on common misconceptions are particularly valuable. Demonstrating that the preconceived notions are incorrect provides an ideal preface to a discussion of the correct physics. An exposition of the correct physics that does not start from students' ideas commands less attention and is less effective.

The multiple-choice question may also serve to reinforce the concepts involved in the previous questions. On occasion, it prompts students to revisit the previous questions with new insight. As an example, consider these estimation and multiple-choice questions from a WarmUp on one-dimensional kinematics.

The estimation question is:
A basketball player jumps 1 meter high off the ground, turns around, and starts back down. Estimate the time she is within 30 cm of the top of her trajectory (her hang time).

The multiple-choice question is:
It is possible for an object to have
a. zero velocity and zero acceleration.
b. nonzero velocity and zero acceleration.
c. zero velocity and nonzero acceleration.
d. Any of the above are possible.

One student made this connection:
"I liked the third question. At first glance I thought only the first two were true and an object could not have zero v and a nonzero a. However, I remembered that in the problem with hang time that for an instant the velocity is zero and the accelleration of gravity is still acting on it. It's not like the object stopped decellerating at $v = 0$, it was decellerating from the moment it left the ground!"

The WarmUp: A Detailed Example

This section presents an example WarmUp for rotational kinematics and dynamics.* For each of the three questions, we provide comments and notes about the question, a few representative student responses, and a description of the classroom session that follows.

Figure 6.1 Images such as this one usually accompany the text of our WarmUps and Puzzles. Such images reinforce the connection to the real-world scenario.

The Essay Question:

"In rewinding an audio or video tape, why does the tape wind up faster at the end than at the beginning?"

Representative student responses:

1. Because the gear in the rewinder is having to turn less amount of weight at the end.
2. Because the tape has a much larger diameter to wrap around. This causes the tape to wind faster.
3. It winds up faster at the end, because it has lesser and lesser mass to be rotated from the other side. Thus, as it winds up, the force from the opposite way constantly decreases. The spool that the tape is winding onto has a much smaller radius than the spool full of tape so it has to rotate several more times than the other spool to wind up the same amount of tape, once you go past equal radii of the spools and get near the end the opposite is true because the right spool is rotating the same amount but the other spool has to rotate much faster to keep up, therefore faster winding.

* Professors and students using Giancoli, *Physics 5/e* will find other questions on this topic at http://www.prenhall.com/giancoli by selecting the warm-ups mudule in Chapter 8.

4. Because the tension is less due to less tape and therefore less resistance to turn.
5. The VCR will make the "tape spools" rotate at a constant speed, given them a constant angular velocity. When you start rewinding the tape, the majority of the tape as on the "source" spool waiting to be transferred to the "target" spool. Initially the target spool has a small radius (due to no or little tape being on it). This means the tape will have a high linear velocity, making it rewind faster. As more tape gets transferred to the target spool, the radius of the target spool gets larger, decreasing the linear velocity, making it rewind slower.
6. The tape winds up faster at the end then at the beginning because it requires more tape per revolution, during the same time frame, at the end then at the beginning.
7. As the tape narrows the end, the diameter of the side it is taking tape from becomes smaller and smaller. Since the tape is moving at constant *V*, the RPM must be greater to get the same amount of tape off the reel.
8. Just like with the skater [in our book], as the tape pulls in closer to the rim, the rotational inertia decreases so the angular speed (tape winding) increases.
9. When the tape gets big on the other side the velocity is greater in turn pulling the tape faster from the small end.

Consider the questions a student faces in analyzing and modeling this situation. There are two spools that are spinning (same rate or different rate?), and the tape is winding on one while it is unwinding on the other. Does the term "faster" in the question refer to the spools, the tape, or both? The students have to try to relate the idea of "winding the tape" to one or more technical terms in the chapter. Students may notice the relationship of "winding" to "rolling without slipping." This illustrates the extendibility built into many of the WarmUp questions.

To answer the question, students must find relationships among the quantities they have identified. They typically draw these from statements they have been told to believe or from inferences they draw from their own experiences. For instance, many students have heard sportscasters discuss the rotation speed of ice skaters, divers, etc. The conclusion is "smaller radii imply higher rotation speeds." The eighth response above demonstrates this. They also tap technical knowledge gained from previous courses or from the textbook.

To prepare for the interactive lecture session, we collect the student responses, read them, and group them into categories. Some categories are general, and some are subject-specific. General categories include responses from students who don't know where to begin, those who misunderstand the question or deliberately answer a different question, those who correctly describe the scenario but never answer the question, and those who bluff their way through the answer. Since responses in these categories are generally not of interest to the entire class, we usually try to deal with them individually, such as through a face-to-face discussion or e-mail exchange with the student. Occasionally, however, a creative misinterpretation of the question can be the starting point of an interesting class discussion.

Some subject-specific categories:

- Focus on dynamics rather than kinematics.
- Focus on the difference between the angular speeds.
- Focus on the linear speed of the tape.

Classroom discussion

We start the session by presenting a handful of representative student responses. The responses are either projected on a screen or read to the class. We try to cycle through the roster so that each student has a chance to see his/her response several times during the semester. One of the objectives here is to foster good technical communication. We take some time to talk about the style, particularly praising the responses that are clear, concise, and complete. Of the responses above, the sixth is a good example. The answer could benefit from more detail, but the statements are concise and unambiguous.

Next we introduce the physics language, including technical terms and symbols. This is accomplished in the context of the student responses. From this example we develop the ideas of angular displacement (θ), angular velocity (ω), and angular acceleration (α) based on the responses. For example, we replace "the RPM" with ω and "winds up faster" with α. We quickly go through this exercise for each of the selected responses.

We now take a closer look at the question. Using lay terminology, we pin down the unstated assumptions that are needed in order to answer the question. This is done as a class discussion rather than a lecture. In this question we assume that the take-up spool is rotated by a motor turning at constant ω. The other spool is pulled by the tape. To keep the tape taut, the "free" reel is acted on by a slight retarding torque (friction) to keep the reel from spinning. We take this opportunity to discuss the relationship between physics, engineering, and the "real world." We point out the necessity of listing the implicit assumptions and their relative importance. Even though this is an introductory course, it is not too early to start the students thinking about estimating sizes and making approximations.

We have set the context for the physics theory. We now present a minilecture focusing on the sticky points: aspects typically missed or misunderstood, mathematical subtleties, and specifics brought up by the class.

We now return to the WarmUp question. This is the time to point out that response number seven expresses the relationship between tape motion and reel motion. We rephrase this mathematically: $v = \omega r$. From this point, we can naturally move to linking the reels through the motion of the tape; we would refer to the correct aspects of responses number three and five. From this connection, we can derive the actual formula, $\omega_1 = \omega_2(r_2/r_1)$. We conclude the discussion by encouraging the students to rephrase this result in English: "The tape winds up faster at the end because the radius of the tape around the take-up reel increases. The angular velocity of the other spool has to accommodate according to the ratio of the radii."

Notice that this discussion contains all of the physics captured in the three-step derivation:

$$v = \omega r \text{ and } v_1 = v_2 \Rightarrow \omega_1 r_1 = \omega_2 r_2 \Rightarrow \omega_1 = \omega_2(r_2/r_1)$$

Obviously, the discussion takes longer than the Spartan derivation, but the benefits of the long version are well worth the time. Explicitly linking the concepts to a real example improves conceptual understanding. We have also taught students how to extract the relevant physics problem from an ill-defined situation, and we continue to build students' communication skills. However, the primary benefit may be improved retention. Students have far better recall of the physics when it is taught in connection with a real scenario that they have considered in preparation for the class.

Extensions

To further exploit this approach, we often ask, "How could we extend this Warm-Up question?" For example, we can discuss the present setup under different conditions or we can analyze different scenarios that involve the same physics. We might ask, "Can you think of a case when the tangential velocity stays constant, and if so, what purpose would that serve?" (e.g., "play" mode). Or, we might consider the bicycle. "What can one say about angular and tangential velocities of the pedal wheel and the bicycle wheel? The sprocket wheel and the bicycle wheel? Why do we have gears?" A classroom discussion of the bicycle frequently leads to the power relation P = (torque)*(omega), and the question "How fast can you pedal up a hill?"

Most of the time the discussion can anticipate future topics. The tape cassette example presented here is a natural lead-in to combining linear and rotational motions: rolling without slipping, center-of-mass motion, etc.

The Estimation Question:

"Estimate the magnitude of the tangential velocity of an object in Indianapolis, due to the rotation of the Earth."

This is a relatively simple question but we usually get a wide range of responses:

1. "I don't know where to start."
2. uses ωr but cannot find the appropriate omega
3. uses ωr but cannot find the appropriate r (typically uses the radius of the Earth)
4. correct and complete

To answer the estimation question the student must:

- Find the relevant physics.
- Supply the missing data.
- Obtain an answer.
- Check for reasonableness.

In our experience, students generally approach this problem in one of two ways. Either they first look for a formula for tangential velocity and then try to apply it to the problem, or they look at the geometry of the motion first to see if that suggests a formula. In each case, they are led to the relationship $v_T = \omega r$, at which point they need to determine the appropriate values of ω and r. These are objective parameters that can be found or calculated from data in reference materials. While ω is the easier of the two parameters to determine, it still poses difficulties to many of our students who do not realize that all points on the Earth have the same angular velocity. Some of the ones who do are unable to actually calculate its value. For many, determining the appropriate value to use for r is even more difficult. The students are expected to state any values they assume or estimate.

Once the students obtain an answer to the question, they often comment on their answer. Many of our students are surprised by how fast points on the Earth are actually moving and question the correctness of their calculations.

Classroom discussion

In the classroom discussion of the estimation question, we retrace the steps we expected the students to take in answering the question. We have developed the physics theory during our discussion of the essay question. Thus, we can use the estimation question to focus on problem-solving strategies, from preparing the scenario for mathematical analysis to checking the answer for reasonableness. Discussing the reasonableness of the answer gives us another chance to extend the question to related topics. Possibilities include: estimating v from $2\pi r/T$, comparing v to the orbital speed of the Earth around the Sun, comparing the magnitude of the centripetal force to mg (to see the relative magnitudes of the quantities), and anticipating satellite orbit speeds and escape velocities.

This is neither the main nor the only classroom session devoted to problem solving. As described in Section One, we explicitly address problem solving in our recitation session. Students work on problems collaboratively in teams with instructors acting as facilitators and troubleshooters. In preparation for class, students hand in (on paper) three or four worked-out problems for credit and comments. In class the focus is primarily on setting up a problem and developing a solution strategy rather than on the mechanical workout.

Variations and extensions

This is an example of a question that lends itself to variations by using different cueing system tactics [Jonassen and Grabowski, 1993, p. 22]. Examples of cueing include graphic cues (see Fig. 6.2), special information (e.g., a table including the Earth's radius and latitudes of major cities), and verbal cues (e.g., use the phrase "linear velocity" rather than "tangential velocity" in the question statement).

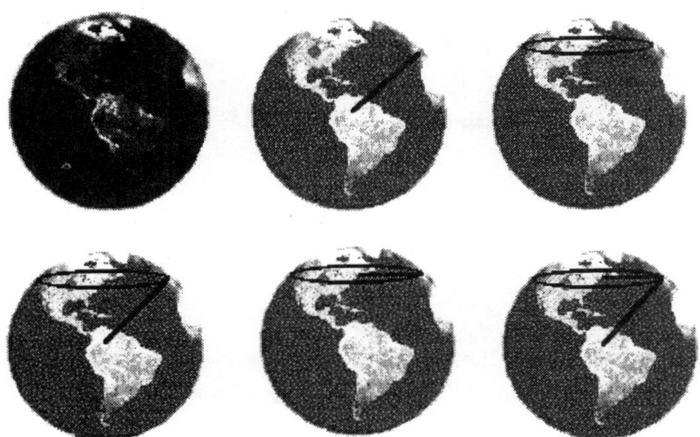

Figure 6.2 Figures such as this provide a graphic cue to students working on the estimation question discussed in this section.

The Multiple-Choice Question:

"The angular velocity of the Earth revolving around the Sun is approximately

a. 365 rad/day c. 12 rad/yr
b. 0.017 rad/day d. 1.0 rad/yr

This question reinforces the concept of angular velocity. Although this question seems like a straightforward application of the definition, in our experience, students have considerable difficulty with it. Part of the problem seems to be that many students do not associate the motion of the Earth around the Sun with the concept of rotation as defined in their textbooks. The question also provides an opportunity to revisit the units used to describe angular quantities.

Classroom discussion

During the classroom time devoted to this question, we discuss the conditions under which it is appropriate to use angular quantities to describe the motion of the Earth around the Sun. Elliptical orbits provide a natural extension opportunity. We refer back to the estimation question in this WarmUp and relate the kinematics of the Earth's revolution around the Sun to the kinematics of its rotation about its own axis.

The Puzzle: A Detailed Example

The Puzzle is a weekly Web assignment designed to integrate the material learned in the preceding week. The classroom discussion of the Puzzle serves as a review

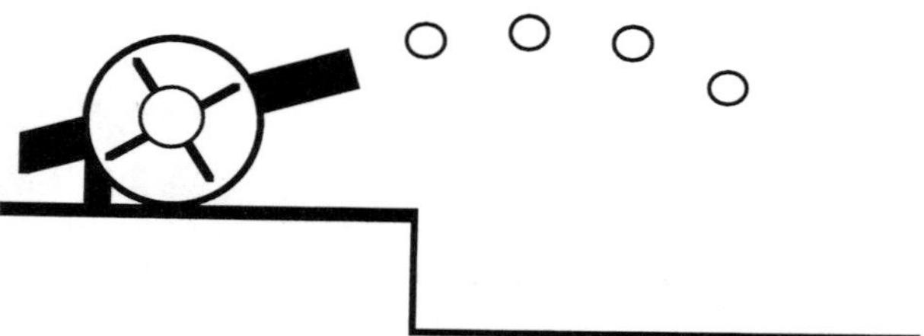

Figure 6.3 The image that accompanies the cannon puzzle.

of the material. It also gives the instructor another opportunity to address problem-solving strategies. The example presented here comes at the end of kinematics.*

The Puzzle Statement:

"You probably know by now, from reading the book and from working the projectile motion problems, that the maximum range for a projectile is achieved when the projectile is fired at 45 degrees. This is true if the launch starts and ends at the same altitude. What if the target is at a lower elevation, as shown above? Is the optimal angle still 45 degrees? If not, is it more? Less?"

Representative student responses:

1. If the spot is lower than the launch spot the cannon should be shot straight out at 0 degrees to get the maximum range.
2. Yes, a 45 degree launch still yields a maximum range. Since the landing spot is lower then the launch spot you need to first find the time the projectile is in the air. Next you use that time to find the distance (x) it travels. Since the final equations involve both sin and cos then to get the max. distance the sin and cos of an angle must be as close to 1 as possible. 45 degrees turns out to be the best angle.
3. A 45 degree launch does not yield maximum range. The launch angle must be changed to the angle that would occur at that height as if it had been launched at the same height as the landing spot. This is because launching it at 45 degrees at that point would be perfect if they were the same height, which is the exact same reason you would have to adjust the angle to the point of the trajectory curve that would correspond to a launch of the lower height.
4. If the landing spot is lower than the launch spot, the situation that is likely to happen is, the projectile will land on a new landing spot that is further than the landing spot with the same height. The reason is, if we take a horizontal line from the launch spot, there will be some point or landing spot where the rule of

* This question is in the warm-ups module in Chapter 3 of the Giancoli *Physics 5/e* companion Web site at http://www.prenhall.com/giancoli.

45 degrees above can be applied. Then if we pull the point to the lower landing line, we can see that the new landing point is further than the original point. So, a 45 degree launch will not yield maximum range, and therefore, the launch angle must be changed to slightly higher.

5. If the landing spot is lower than the launch spot, 45 degrees is no longer the angle for the maximum range. The new angle for the maximum range would be some angle less than 45 degrees depending on the height of the extra distance. This is true because as the landing site becomes closer than the level of launch, the two parabolas (one at 45 degrees and one < 45 degrees) cross each other. When this happens the parabola with the smaller angle widens faster than the 45 degree angle parabola, which increases the range to a greater distance for the smaller angle. Comment: These are very cool problems! I had a little trouble with the reasoning for question 1, and could only come up with a qualitative description to explain my answer. Can you please give the quantitative or mathematical reasoning for this question in class?

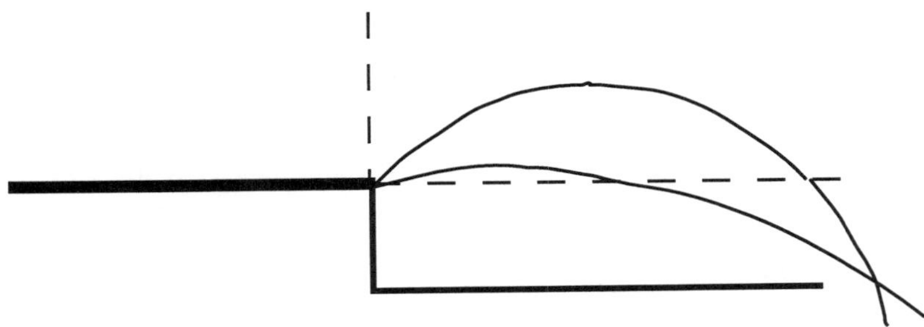

Figure 6.4 The image constucted during classroom discussion of the cannon puzzle.

Classroom discussion

We usually start the puzzle discussion by displaying the most correct and complete student response. In this case that would be response number five. The response is based on an analysis of the trajectories, so the classroom discussion would start there. With the help of the author of the response, we would construct the trajectories shown in Fig. 6.4

We would then carefully reproduce the student's reasoning and his/her conclusion that for a target at a lower elevation, a reduced angle would give a longer range. Next we would develop the "range" formula for this case. This is a nice opportunity to review the kinematics equations. Analyzing the expression, the students will agree that to obtain an analytical answer to the question, we need to take the derivative of the range expression with respect to the launch angle, set the resulting expression to zero, and solve the resulting equation for the launch angle. Two facts emerge:

1. The resulting algebra is pretty ugly.
2. The solution depends on the difference in elevation.

Now we have three options:

1. Leave it at that.
2. Write down the solution and test the limit of zero height difference.
3. Invite the students to do the algebra for extra credit.

Which option we take will depend on the class. If only one or two students are capable of option 3, it is probably best to take option 2. Notice that the student who submitted response 5 is likely to take option 3 given some incentive and probably some out-of-class help.

Extending the puzzle

The puzzles are usually extendible. We encourage the students to think of these extensions, and we bring them up during the classroom discussion of the puzzles. The puzzle presented here has several obvious extensions:

1. Analyze the problem for a target at a higher elevation.
2. Analyze the problem of shooting up an incline. (This requires a redefinition of "range," which most students don't find obvious.)
3. Analyze the problem if air resistance cannot be neglected. This is an opportunity to bring in computer modeling.

Of course, all of these features are present in the behind-the-chapter problem collection, and they are usually mentioned in the traditional lectures and recitations. Still, students are more likely to attend to these subtleties and to retain them if they are presented in the Just-in-Time setting.

What Do Students Think?

At the USAFA:

> The Preflights are a very good idea because they are a very good preparation for the following day's lesson. They should continue to be graded so that it "forces" the student to do them the night before the lesson. It also gives the student the opportunity to ask any questions that the instructor can prepare the answer to before class that day.
>
> ... Yes, [the Preflights are] a very good idea for a number of reasons. First, they get us to review and at least skim the material to be covered in the next lesson's class .... It forces me to make the class a priority the night before. I know they are geared for about 10–15 minutes, but sometimes I spend a lot longer than that on them, but I think in the long run, the time is well spent ....

At IUPUI:

I do believe puzzles are a good idea because it makes the student think about physics. Thinking is the most important thing needed on a test and I believe the puzzles promote it.

At first glance, some of the puzzles looked impossible but usually something would click and I could solve it. I am biased towards them because they are a way to make up points that I lose to test anxiety on the 'real' exams. Puzzles are done without a clock running, and they are actually good practice for the final because I had to step back and think about what was really going on in simpler terms.

Chapter 7: The Insightful Lecture

by Col. Rolf Enger, USAFA

Teaching is a challenging occupation. Teachers today must be subject-matter experts, curriculum designers and developers, cheerleaders, evaluators, and even mindreaders. Of course, there was a time when teachers were thought to be primarily subject-matter experts who "professed" the knowledge they possessed to attentive students eager to absorb their every word. However, not so today! Although the "professorial" method is still practiced on many campuses, a growing body of teachers are recognizing the value of actively engaging students in a discovery-oriented learning experience in the classroom itself. For example, Eric Mazur at Harvard University has perfected a technique he calls Peer Instruction [Mazur, 1996]. During class, Mazur confronts students with a series of carefully constructed questions designed to highlight typical student misunderstandings and emphasize key ideas in the day's lesson. Students discuss the answers with each other and respond either on paper or electronically. Mazur collects the students' responses and instantaneously modifies his subsequent lecture based on these replies. Other teachers have embraced more elaborate collaborative learning strategies, organizing students in teams and giving them rigorous exercises to complete in and out of class. In some applications, teachers assign specific roles to each student, further encouraging them to rely on each other to arrive at a solution.

The "insightful" interactive lecture is designed to work in tandem with collaborative peer learning and other interactive classroom strategies. In particular, it is designed to reduce the need for teachers to "read the minds" of their students, especially at the beginning of class. All teachers have experienced the uncertainty that comes from walking into class with little or no information about the preparation level of their students. Did the students complete the assigned homework? Did they understand the textbook? Could they successfully complete the assigned homework problems? In a traditional classroom, the teacher will attempt to answer these questions by asking a few probing questions, watching facial expressions as students attempt to answer these questions, perhaps giving a short quiz or even giving students an in-class exercise to complete. But when using the insightful lecture strategy, that same teacher can arrive at class already knowing the answer to each of these questions for all or most of the students. What a difference that makes for both the teacher and the students.

Constructing the Preflights

Teachers begin by detailing the learning objectives for the lesson and identifying the parts of each that can reasonably be accomplished by the students prior to class. Appropriate readings, questions, and homework problems are assigned, each designed to help the student develop the conceptual understanding and operational skills needed for the upcoming class. The teacher designs the WarmUp or Preflight for each lesson to gather data about the success students have had in acquiring the knowledge and skills deemed as "prerequisite" to the class meeting.

The insightful Preflight typically contains four to five questions, each based on the lesson's learning objectives. We have found these questions most insightful when they are designed to elicit successful responses from students. In writing each question, we ask ourselves, "What do we expect students to learn on their own from the assigned readings, questions, and homework problems?" Since the Preflight questions are written based on the answer to this self-inquiry, the expectation is that all students will successfully complete each Preflight question. Thus, the most powerful insight occurs when students fail to achieve this level of success. It's worth pointing out here that we try hard to ask questions that we expect students to answer successfully. It is of little use to know that the students cannot successfully answer any questions. Since learning occurs best when new knowledge or skills can be built upon previous learning, it is important to design questions that identify the extent (as well as limitations) of the students' current understanding.

As with the "mindful" strategy, Preflights are included as a component of the students' homework assignment. Students typically complete them the night before class and submit them via the World Wide Web. Teachers read and react to the student responses prior to coming to class.

Preflights as Filters

The Preflights can be thought of as a filter. Prior to reading the responses, perhaps even prior to writing the Preflight, the teacher has selected a series of possible classroom activities such as minilectures, short videos, demonstrations, in-class exercises, or even a lab. The Preflight responses then serve as a filter. When gaps in student understanding or ability are detected by the Preflight, a classroom activity addressing that issue "passes through" the Preflight filter and is ready for use to remediate that difficulty during the subsequent class.

It is worth reemphasizing that the most meaningful insight in the insightful strategy occurs when students fail to successfully answer a Preflight question that the teacher expected the students to answer correctly (by completing the assigned homework). When this happens, it represents a failure to accomplish an objective for the lesson *and* it informs the teacher that further work on this concept or skill is needed. By receiving the Preflight information prior to class, the teacher knows how prepared the students are before the class ever begins. Class then becomes more effective and efficient, building on the knowledge and skill base the students have demonstrated in the Preflight.

Classroom Discussion

The in-class experience for both teachers and students when using the insightful strategy is similar to that described above for the mindful strategy. We typically discuss each question, even those most students answer correctly. This reinforces the correct response and gives encouraging feedback to those who have answered correctly. We spend more time working with questions that students answered incorrectly, employing one or more of the prepared strategies we brought to class (short lecture, demo, video clip, etc.). We also build on the Preflight, using that confirmed base of knowledge as the foundation for further exploration. Collaborative techniques such as those used by Mazur and others can easily be employed. In fact, the classroom session is very similar to that found in a traditional course, although the outcome can be superior. Since student submissions shape lesson design, students are more motivated, tend to pay more attention, and gain deeper knowledge.

The following is an example of a typical Preflight designed in accordance with the insightful lecture implementation. The topic is RC circuits (circuits that contain both a resistor and a capacitor). This material is typically taught in the second semester of an undergraduate introductory physics course.

The Learning Objectives:

- ❑ To understand how and why the current in an RC circuit changes with time.
- ❑ To calculate the current in an RC circuit as a function of time.
- ❑ To calculate the charge in a capacitor as a function of time.
- ❑ To calculate the time constant for an RC circuit.
- ❑ To understand the physical significance of the time constant in an RC circuit.
- ❑ To calculate the voltage across a capacitor as a function of time.
- ❑ To calculate the time constant for a known resistor-capacitor circuit.
- ❑ To determine the capacitance of an unknown capacitor.

The Preflight:

1. An RC circuit is a circuit that contains both a resistor and a capacitor. Suppose such a circuit also contains a battery and a switch as shown in the figure. When the switch is closed, why does the current in the RC circuit change with time?
2. What equation from the book would you use to calculate the current in an RC circuit as a function of time?
   a. Give the equation number from the book.
   b. Indicate what each letter in this equation stands for.

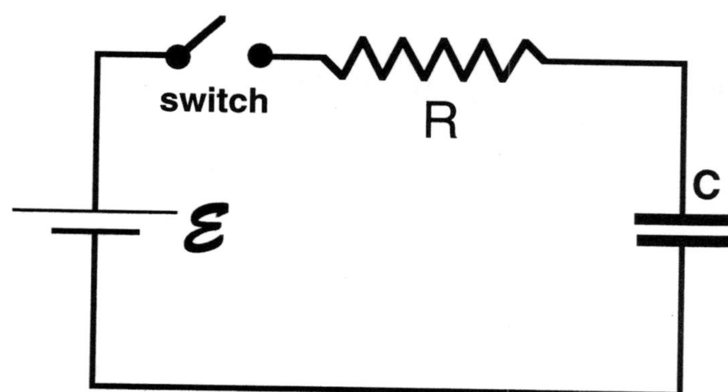

Figure 7.1 The diagram that accompanys the RC circuits Preflight.

3. An RC circuit consists of a 12 V battery connected in series to a 25 Ω resistor and a 150 mF capacitor. What is the time constant for this circuit?
4. The electronics in televisions and computer monitors contain large capacitors that help provide the high voltage needed for operation. Using your knowledge of RC circuits, why is it that the following warning can be found on the back of such devices: "CAUTION: To prevent electric shock, do not remove back." (Hint: Why doesn't it just say "Be sure to turn off and unplug unit before removing back"?)
5. When the switch in an RC circuit containing a 12 Volt battery, a 4 Ω resistor, and a capacitor in series is closed,
   a. the initial current through the resistor will be 3A and then diminish to zero.
   b. the initial current through the resistor will be zero and it will grow to 3A.
   c. the current through the resistor will stop when the capacitor is discharged.
   d. Both b and c are correct.

Analysis of Student Responses

As expected, there was a range of student success on this Preflight. Some gave solid answers. Consider, for example, this response to the first question from one student:

> Immediately after the switch is closed, current flows rapidly through the resistor to the capacitor. The capacitor begins to accumulate charge rapidly, and as the capacitor voltage approaches the battery voltage, the resistor voltage must drop as well. If the resistor voltage is to drop, then current must drop also because

> $V = IR$. Now, since current drops along the resistor, the capacitor charges more slowly since less charges are reaching it.

Others came close:

> The current changes with time because when the current starts to flow, charge accumulates on the capacitor at an ever-decreasing rate. This occurs because as the capacitor voltage rises, the resistor voltage decreases, so the current decreases at an ever-decreasing rate.

Others clearly had misconceptions about how currents and potential interact in a circuit:

> The current will take time to get past the resistor and also to fill the capacitor. When it is not involved with these things, it will move normally.

And still others gained enough insight to realize that they had "no idea" [emphasis is the student's]:

> Capacitor voltage approaches battery voltage.... The current through the resistor or charge of the capacitor becomes very small going closer and closer to zero. I DON'T EXACTLY UNDERSTAND WHY...

The above is a student who now recognized that he had a need to know more. Another student wrote, "Can we do some example problems? I'm kind of confused from the reading for this section." Still another wrote, "I would like to get some EI (extra instruction) this week." These are students who have gained important personal insight in completing the Preflight and have a heightened motivation to learn.

An Insight Gained

The above Preflight proved to be very insightful for faculty, too. For example, although experienced teachers expected question 3 to be trivially easy, only 10 % of the students answered it correctly. The solution simply requires a student to multiply R and C together (since the time constant $\tau = RC$). In reading a variety of student responses, it quickly became clear that the concept of a "time constant" had no meaning for students. In fact, although some could correctly identify RC as the time constant (a part of the answer to question 2), they were unwilling to do the simple multiplication to calculate it. For many, the inclusion of the voltage in question number 3 was troubling. They were convinced that the solution required the use of that piece of data. For nearly all, it was clear that the time constant was a mysterious entity they simply did not understand.

The students' difficulties with question 3 had a profound impact on at least one experienced faculty member at the academy. Having taught RC circuits for several years without observing any apparent student difficulty with the "time constant," this teacher had relegated the "time constant" to a few short minutes at the end of the lesson—more as an aside than as a main concept. However, when confronted with the student Preflight data, this teacher quickly inverted his lesson plan and made the time constant a featured focus of the lesson. The Preflight had indeed provided a much needed and valuable insight.

What Do Students Think?

In wrapping up this section on the insightful lecture implementation, we include a few comments from students, collected from an anonymous survey about these Preflights. The students were responding to the question, "Do you think Preflights are a good idea?" Note that the students recognized that Preflights, when used as part of the insightful lecture implementation, provide teachers and students with both motivation to learn and important insights about the quality and amount of that learning.

In these first two comments, the students are reflecting on the insight that comes from preparing for class, something that is motivated by the Preflight assignment.

> Preflights are good because they help us think and give us motivation to at least read the assigned homework for the next lesson, instead of waiting until class to hear about what we are to learn.

> Yes, they really do get me thinking of the material prior to class so that I am not clueless and starting from scratch.

In these final two comments from students, the students reflect on the insight that comes from recognizing that they have a "need to know" more:

> Yes [Preflights are a good idea]. So you can see what is important for the lesson. Also, it helps the instructor see what you do not understand.

> I think the Preflights are useful for exposing to us what we understand or need more explanation on.

Section Three: Resources

Chapter 8: WarmUps

Chapter 9: Puzzles

Chapter 10: Physlet Problems

Chapter 11: Scripting Physlets

Chapter 12: Communication

Chapter 13: Frequently Asked Questions

Copyright and license notices for WarmUps, Puzzles, Physlets and Physlet Problems:

Many of the WarmUps, Puzzles and Physlet problems discussed in this Section are on Prentice Hall's companion Web site for Giancoli, *Physics 5/e* and will be on the Web sites for Giancoli, *Physics for Scientists and Engineers 3/e* and Wilson/Buffa, *College Physics 4/e*. Those sites will be periodically updated with new questions. A corresponding correlation key between the questions discussed here and those on the Prentice Hall companion Web sites will be maintained on Prentice Hall's web catalogue page for this book at

http:// www.prenhall.com/allbooks/esm_0130850349.html.

Chapter 8: WarmUps

General Principles

We have several basic objectives in constructing a WarmUp assignment.

- Set the framework for the physics to be studied in everyday terms.
- Pique the student's curiosity; generate a need to know.
- Relate material to be learned to material already studied.
- Give the student a sense of confidence.

Having done the assignment, the student should be able to say, "I know what the lesson will be about. I understand the scenario, and I understand the question. I have some idea of the answer, but I am not sure and I don't know all the details."

To accomplish the above objectives, we present each question in a context that can be understood without reading the textbook. However, the question should incorporate the important concepts from the chapter. We typically present a scenario and with one or more questions ask the student to predict the outcome. It is important that the question be simple and straightforward so that the student has no trouble understanding what is being asked. It is equally important that the question be complex enough so that some conceptual understanding of physics is necessary to provide a complete answer. When the underlying physics contains frequently held erroneous preconceptions, the WarmUp question is constructed so as to flush out these preconceptions. The classroom discussion of such questions must be explicit enough to persuade the students to give up the preconceived notions. It may well be necessary to bolster the case with a well-planned demonstration or activity. Where possible, we leave some particulars out, to be supplied by the student. This adds variety, which makes the classroom discussion more interesting.

It is important to make the questions good discussion starters. It is not uncommon for students in JiTT courses to be talking about WarmUp questions in their conversations before class begins. The best WarmUp questions engage the students enough that they are actually thinking about the questions and their answers prior to class. Good questions fall into different categories but generally share one or more of the following features.

- ❑ attention catchers
- ❑ refer to current events
- ❑ are just ambiguous enough to cause the students to really pause and think
- ❑ explicitly confront problem areas identified by physics education research
- ❑ pick up on a thread from a very recent class discussion or activity
- ❑ are not too easy or too hard
- ❑ include at least some (preferably all) nontechnical phrasing

Another feature that often works well to generate discussion and facilitate understanding deals with multiple-choice questions. We sometimes include more than one choice that could be correct, depending on the assumptions the student makes about the setup. These questions then are deliberately ambiguous or loosely defined. It is not a good idea to have this kind of question unless the students are explicitly told, "There may be no one correct answer to this" and "The purpose of this question is to elicit discussion in class." Furthermore, students should not be awarded course points based on correctness in such cases of deliberately vague or ambiguous problems, for they can easily become disheartened. The students should not feel that they are being penalized for having been "tricked" but rather should feel that the question has brought about a "teachable moment" that has helped them to solidify their understanding.

Mechanics

1a. During aerobic exercise, people often suffer injuries to knees and other joints due to high accelerations. When do these high accelerations occur?

The short answer is that high accelerations occur when the person's feet hit the floor. This question prepares the student for a discussion of definitions of kinematics quantities. Many students find it difficult to grasp the idea of rate of change. This example involves abrupt stops (high rate of change of magnitude), abrupt change in direction (calling attention to the vector nature of kinematics quantities), and—implicitly at least—it prepares the ground for associating accelerations with forces.

1b. Estimate the acceleration you subject yourself to if you walk into a brick wall at normal walking speed. (Make a reasonable estimate of your speed and of the time it takes you to come to a stop.)

This can be answered by estimating the time it takes to stop you or by estimating the distance the body moves as you compress against the wall. The second estimate is easier and more intuitive, but it does not use $a = dv/dt$, which, in our experience, is what most students use at this point.

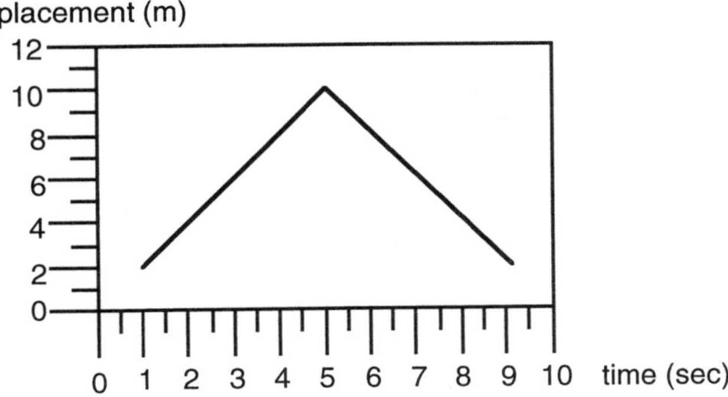

1c. The figure shows a displacement vs. time graph for a ball rolling along the floor. What was the speed of the ball during the third second?

a) 1 m/s
b) 2 m/s
c) 3 m/s
d) It cannot be determined from the graph.

This question is preparation for discussing kinematics graphs. The answer, of course, is 2 m/s. We discuss the unrealistic situation at $t = 6$ sec, let the class decide what is wrong with that point (No derivative exists. What does that mean physically? What is the graph trying to show? How can we make it right?). From there we go on to graph v vs. t, a vs. t, and have a general discussion of motion graphs.

2a. A man drops a baseball from the edge of a roof of a building. At exactly the same time, another man shoots a baseball vertically up toward the man on the roof in such a way that the ball just barely reaches the roof. Does the ball from the roof reach the ground before the ball from the ground reaches the roof, or is it the other way around?

They reach their respective destinations at the same time. Most students get this correct, but their logic is often faulty or incomplete. We want them to answer with a short essay, but we want the verbal supported by an appeal to mathematical arguments.

2b. Estimate the time it takes for a free-fall drop from 10 m height. Also estimate the time a 10-m platform diver would be in the air if he takes off straight up with a vertical speed of 2 m/s (and clears the platform of course!)

Estimated time for the drop is straightforward, about 1.4 sec. The quickest way to estimate the time in the second case is via average velocity, which is about –6 m/s. Having made the estimate, students then check this value by actually solving the quadratic equation.

2c. A football is kicked at ground level in such a way that the horizontal component of its initial velocity is 7 m/s, while the vertical component is 10 m/s. When it returns to the ground, the football will have a speed of about
a) 12 m/s.
b) 10 m/s.
c) 17 m/s.
d) none of above

The final speed is equal to the initial speed (12.2 m/s), but the vertical component of velocity is reversed. This is useful as a classroom activity to put together the 2D kinematics equations.

3a. The word "push" is a reasonably good synonym for the word "force." Is "hold" a synonym for force? How about "support"? Can you think of others?

This vague question is intended to generate a feel for the meaning of the term force. Most physics terms with counterparts in everyday language carry connotations that mask the essential physics. A force will cause an acceleration unless it is balanced out by other forces. Students always pleasantly surprise us with a rich collection of choices (including bad ones, such as "power").

3b. The engine on a fighter airplane can exert a force of 105,840 N (24,000 pounds). The take-off mass of the fighter plane is 16,875 kg (it weighs 37,500 pounds). If you mounted the airplane engine on your car, what acceleration would you get? (Use a reasonable estimate for the mass of your car.)

This provides a good introduction to a discussion of mass vs. weight, kilograms, newtons, pounds, slugs, and other units. It is also good for giving students a feel for the size of these units. A 1000 kg car acted upon by a 100,000 N force would accelerate at about 10 g's.

3c. A locomotive pulling a train is accelerating the train on level track. The tension in the couplings is
a) the same anywhere along the train.
b) least between the locomotive and the first car.
c) least between the last car and the caboose.
d) none of the above

This WarmUp introduces free-body diagrams. It is deliberately vague. We do not know how many cars there are. We don't know the masses of the cars. Yet we can still conclude that the correct answer is c. It is also worthwhile to refer back to this question when discussing the (constant) tension in a massless string.

4a. One farmer is bragging about his strong horse, while his neighbor is proud of her powerful horse. If the farmers are giving the words "strong" and "powerful" their correct physical meanings, what can you say about the two horses?

This question can be used to start developing the notions of work, energy, and power and relating them to the notion of force. It can be extended to car engines, athletic performance (e.g., weightlifting versus sprinting versus marathon running). The first horse could be one that can carry a heavy load but is very slow, while the second horse might be fast and nimble but not necessarily very strong.

4b. A good professional baseball pitcher throws a ball straight up in the air. Estimate how high the ball will go. (A good throw can reach 90 mph.)

This is a straightforward application of the work-energy theorem (restricted to kinetic energy), which we use to introduce the notion of potential energy. Ignoring air drag, we get an unrealistically high estimate (80 m). Some class time can now be set aside to talk about realistic estimates, perhaps using a spreadsheet or some other software to construct a more realistic model of this case.

4c. An object is moving in such a way that the rate at which it is losing gravitational potential energy is the same as the rate at which it is gaining kinetic energy. You can conclude that
a) the object is falling freely.
b) there is a net force on the object.
c) both a and b
d) neither a nor b

If the object is gaining kinetic energy, it has nonzero acceleration; therefore, it is subjected to a net force. It need not be in free fall. It could be sliding down a frictionless ramp or swinging on a string (ignoring friction).

5a. Consider a hand holding a ball at rest. The ball weighs 10 N. The forces acting on the ball are its weight and the force of the hand on the ball. The forces acting on the hand are the weight of the hand and the force exerted by the ball on the hand.
How big is the reaction force associated with the weight of the ball? What object is responsible for that force?
How big is the force that the ball exerts on the hand? How big is the reaction force associated with this force?
Now let the hand exert a 15 N force on the ball.
How big is the reaction force associated with the weight of the ball? What object is responsible for that force?
How big is the force that the ball exerts on the hand? How big is the reaction force associated with this force?

The third law is elusive. It is also nonintuitive. Nevertheless, it is an important underpinning to many arguments students will encounter in a first-year course. In one form or another, this question appears in most introductory physics books. Our students have trouble with this question, particularly with the second part. The re-

action force that pairs with the weight is the force that the ball exerts on the Earth. It is 10 N in both parts of the question. The force that ball exerts on the hand is 10 N in the first case and 15 N in the second case.

5b. Suppose the astronaut working on the *Hubble* telescope got himself completely loose of the shuttle and threw away a 2-kg hammer at 2 m/s. Could the astronaut remain at rest? If not, how fast would he be moving? How could he stop himself from drifting away?

In our experience, this is an easy question. It is intuitively clear to most students that the astronaut will recoil. Estimating the astronaut's mass at 80 kg, the recoil speed estimate is 5 cm/s. Without a tether to hold on to the astronaut could throw something else away in the direction of his own drift velocity. Even though this is a simple question, we find it useful to call attention to the vector nature of momentum.

5c. A 500-g piece of clay hits a stone at 80 m/s and sticks to it. The clay momentum must change by
a) 80 Ns.
b) – 40 Ns.
c) 40 Ns.
d) More data is needed to answer.

The answer to this question is d. This is an inelastic collision with the state of motion after collision unknown. Our students typically pick b. Note: Students would justifiably label this a "trick" question if it appeared on a test for credit. It is fair and useful as a WarmUp. With an occasional question like this, students begin to read the questions more carefully and to pay attention to the assumptions they make before they commit to an answer.

6a. A skydiver whose parachute is fully deployed is descending at constant speed. Describe what is happening to her kinetic energy, her potential energy, and her total mechanical energy as she falls. Is any work being done? If yes, where does it go?

Students answer this question in preparation to in-class treatment of the work-energy theorem. We revisit all the definitions. Total work in this case is zero (there is no change in kinetic energy). There is work done by Earth's gravity, but there is also negative work done by the air drag. We point out how work done by the Earth's gravity is expressed in terms of the change in gravitational potential energy.

6b. Some athletes can put out as much as 700 watts in short bursts. If you could sustain such an energy output long enough to reach the top of a mountain, how much time would it take you to get up there from the valley, 7000 feet below the peak?

To hike up 7000 ft is an all-day trip for most people. Hiking at 700 watts, an 80-kg hiker would reach the summit in about 40 minutes. From this we can go on to estimate the average speed (50 m/s straight up, 74 m/s on a 45° slope). Questions like this are useful to relate less familiar physical quantities to more meaningful everyday situations.

6c. A force of 100 N, acting over a distance of 10 m, increases the kinetic energy of a moving object by 500 J. This information is sufficient to conclude that
a) the object was accelerated.
b) there was friction present.
c) the potential energy changed.
d) all of the above

Here is another stab at the work-energy theorem. The change in kinetic energy does not account for the work done by the applied force. So other forces must have been present. What are the possibilities? Conservative forces? Nonconservative forces? The only valid conclusion in the above set is a.

7a. In rewinding an audio or video tape, why does the tape wind up faster at the end than at the beginning?

Note that this question is discussed at length as an example of a WarmUp in Chapter 6. The question is simple enough, yet the number of physical concepts that have to be sorted out is quite large: Linear quantities vs. angular quantities. The relations between the motions of the spools and the tape. Ultimately, the analytical model of a physics problem is a system of equations (algebraic, differential, or integral). It takes time and effort to cultivate the skill to navigate comfortably between the real world (the tape player), the conceptual model, and the symbols in a system of equations. Working thoughtfully through a question like this one builds such insight. While this would be an awful exam question, it is a very good WarmUp exercise.

7b. Estimate the magnitude of the tangential velocity of an object in Indianapolis due to the rotation of the Earth.

This question is also discussed in Chapter 6. For most of our students, visualizing the spatial relations in this question is not easy. It is especially challenging to the students who are still wedded to the idea that somewhere in the book there is a formula to handle questions like this. "I could not find anything in the book that would help me answer this" is not an uncommon comment. See Chapter 6 for possible alternate phrasings and/or hints.

7c. The angular velocity of the Earth revolving around the Sun is approximately
a) 365 rad/day.
b) 0.017 rad/day.

c) 12 rad/yr.
d) 1.0 rad/yr.

This is the third of the three WarmUp questions discussed in Chapter 6. This appears in the set of questions that introduce rotational kinematics. We use this question to firm up the definitions and the units of rotational quantities. Many students do not associate the word "rotation" with the motion of the Earth around the Sun.

8a. Suppose you are standing on the edge of a spinning carousel. You step off, at right angles to the edge. Does this have an effect on the rotational speed of the carousel?
Now consider it the other way. You are standing on the ground next to a spinning carousel and you step onto the platform. Does this have an effect on the rotational speed of the carousel? How is this case different from the previous case?

We use this question to introduce angular momentum. Part one: The answer to the question is "no." Good discussions surround many questions like these: "What happens to the angular momentum of the carousel?" "Was there a net torque exerted on the carousel as you stepped off?" "What happened to your angular momentum?" "How does one define the angular momentum of an object on a spinning platform?" "How is that related to the moment of inertia associated with the object?" "How does one think of angular momentum intuitively?" "How do we use integrals of motion?" "Did any of the student answers mention angular momentum?" and "Did they use the term correctly?"

For part two there is no single answer. The answer depends on your state of motion as you step on. A good discussion starter is "List several possible ways you can accomplish this." Also, list a few student answers to part two. There will be bits and pieces of correct physics, but in general it will be pretty garbled. Here is a chance to start to sort it out. Can the students relate this to what they remember about inelastic collisions?

8b. The mass of the Earth is about 6×10^{24} kg, and its radius is about 6×10^{6} m. Suppose you built a runway along the equator and you lined up a million 10,000-lb. airplanes and had them all take off simultaneously. Estimate the effect that would have on the rotational speed of the Earth. Note: The mass of a 10,000-lb. airplane is about 4500 kg.

What physics do we use? Conservation of angular momentum. To answer, first decide which way they take off, east or west. Next, make a reasonable takeoff speed estimate (order of magnitude ~ 100 mph). Estimate how much angular momentum the planes gain (lose?). How does that affect the Earth? Compare this to shooting a cannon from a boat.

8c. An athlete spinning freely in midair cannot change his
a) angular momentum.
b) moment of inertia.
c) rotational kinetic energy.
d) All of the above conclusions are valid.

The correct answer is a. What would he do to change his kinetic energy? Can he decrease his kinetic energy of rotation? Where does that energy go?

9a. Consider a mass on a spring, oscillating under the influence of a non-conservative retarding force, such as air drag. How would the retarding force affect the period of the oscillations? In a sentence or two, justify your answer.

Most students will guess that the period must decrease. They have seen it happen. What kind of arguments can be given to support that conclusion? An oscillating mass is accelerated (or decelerated) by the forces acting on it. Travel time for any part of the cycle (say, from maximum displacement back to equilibrium) will depend on this acceleration. It is not a simple square-root dependence, as with constant acceleration, but it must still be true that a decrease in the acceleration will lengthen the time. Is it still true that the period is independent of amplitude as it was for a linear restoring force? How can we see all this from the equation of motion? Go as far as the class will follow. Challenge individual students who are more calculus savvy.

9b. When you push a child on a swing, your action is most effective when your pushes are timed to coincide with the natural frequency of the motion. You are swinging a 30-kg child on a swing suspended from 5-m cables.
Estimate the optimum time interval between your pushes.
Repeat the estimate for a 15-kg child.

This is a simple preparatory question before we tackle forced oscillations. It can be disposed of quickly if the syllabus does not emphasize (or even include) a discussion of forced oscillations and resonance. It can be made into a full-blown development if appropriate.

9c. When a simple harmonic oscillator is subject to a damping force
a) the total mechanical energy is conserved.
b) the amplitude is decreasing but the total mechanical energy remains constant.
c) the total energy and the amplitude are decreasing.
d) None of the above is true.

This is related to question 9a, this time from the energy point of view. The discussion can be expanded to include forced oscillations and resonance.

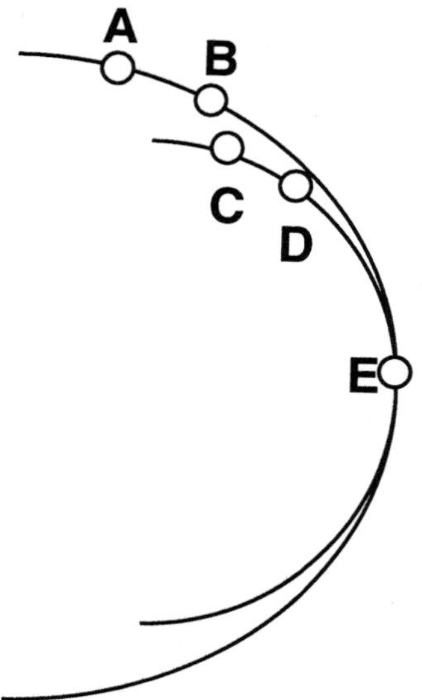

10a. Two different planets are orbiting the same star along two different orbits. Compare the speeds of the planet at points A, B, and E. Compare the speeds of the planet at points C, D, and E.

The planet in the elliptical orbit is to be shifted to the circular orbit as it passes point E. Does it have to speed up or slow down?

Where is the star? What is the direction of the gravitational force on the planet at point A? Does the force have a tangential component? Is there a tangential acceleration? Yes to both questions: The planet is speeding up. What is the direction of the gravitational force on the planet at point C? Does the force have a tangential component? Is there a tangential acceleration? No to both questions: This planet is maintaining constant speed.

The planet that reached point E from point B is moving faster than the planet that came from point D. How do we know that? What would be the correct centripetal force on the planet at point E if it were to stay on the circular orbit? The gravitational force at E does not meet this requirement for the planet that came in from B, or else the planet would follow the circle. Is it moving too fast or too slowly?

10b. The space shuttle has to take a spy satellite to a spot 1200 m above the Earth's surface and release it into a geosynchronous orbit. Is this possible? If yes, how fast must the satellite be moving when it is released?

For many of our students, orbital mechanics clicks for the first time when presented with this kind of question. Most students, except for the very perceptive ones, believe that a satellite can be made to do pretty much anything you want, anywhere you put it. Common student answers to this problem are: 8200 m/s or 436 m/s.

8200m/s is the orbital speed at an elevation of 1200 m (which is an absurd figure anyway). At 436 m/s, the satellite would circle the Earth in 24 hours, but the Earth's gravity could not maintain such motion.

The answer, of course, is: No, this orbit is not consistent with orbital mechanics. Orbital speed and the radius of the orbit are related. They know this, of course, but at this point it's just another formula.

Start the classroom discussion by describing the problem of crowding lots of communication satellites into a limited amount of geosynchronous space. Why is geosynchronous space limited? There is plenty of room around the earth for orbits.

10c. Kepler's third law is a statement of

a) conservation of energy.
b) conservation of angular momentum.
c) both
d) neither

What is Kepler's third law? How does it follow from the laws of mechanics for a circular orbit? How about for any orbit?

Thermodynamics

1a. A piece of wood at 80°C can be picked up comfortably, but a piece of aluminum at the same temperature will give a painful burn. Why is this?

Students often fail to distinguish between temperature and heat. They are also generally unfamiliar with the idea of thermal conductivity. Another unfamiliar notion is that the amount of heat that gets transferred to their hands depends not only on the temperature difference between the hot object and their hands but on the conductivity of the hot object (and of their hands). This question helps clarify the terminology and emphasizes the idea of heat as energy flow.

1b. Estimate the amount of heat that would be necessary to raise your body temperature by 1°F.

Some students who have been exposed to the BTU as a unit of heat or energy will realize that using the BTU can make this estimation easy. Some students may, however, be uncomfortable dealing with the difference between the human body and an equivalent weight of water. Many other students will correctly determine another good starting point for this estimation: The heat necessary is given by the product of the specific heat of the material being heated and the temperature change. Coming up with an estimate for the specific heat of a human body is a major obstacle, as many students will balk at approximating the body as any of the materials whose specific heats are listed in a table in their textbook (like water!). They will typically not try to look up the heat capacity of a human body (83% of the heat capacity of water). Unit conversions can also be a problem; many students will forget that 1°F is not the same as 1°C, or they will convert incorrectly.

1c. A BTU is a unit of
a) energy.
b) thermal conductivity.
c) power.
d) none of the above

Since the phrase "British Thermal Unit" doesn't convey much information that would help someone figure out what a BTU actually is, students need to learn that it is a unit of energy. Discussing the interesting history of its definition is often worthwhile in class. Since some students will have heard of BTUs in the context of heating and cooling systems, they may be inclined to choose (b) because of the word "thermal" or (c) because of the power requirements of such systems.

2a. The ideal gas law says the pressure of a gas on its container walls depends on the temperature of the gas. Please explain why this is.

Many students will essentially restate the formula $PV = nRT$ in words rather than try to offer a physical explanation of pressure. Others may mention that the pressure's dependence on temperature is a consequence of Boyle's law and Charles' law (or the Gay-Lussac law). Some answers will involve a microscopic view of a gas as molecules that have high speeds and that hit against the container walls but fail to make important connections such as the connections between "hit," "force," and "pressure" and between "speed," "kinetic energy," and "temperature." This question provides a good chance to review many fundamental concepts of kinetic theory and to link key ideas to one another.

2b. Estimate the number of gas molecules in one cubic meter of air in the lecture hall on an average day.

Most students suspect that using the ideal gas law is a good way to start this estimation. However, most students prefer the form that they are familiar with from chemistry ($PV = nRT$). Thus, they often do not realize that the number density of gas molecules is readily accesible ($N/V = P/kT$). Other trouble spots are typically associated with using appropriate units (especially temperature in Kelvin) and in unit conversions. Most students are surprised at the very large density. This affords a good opportunity to discuss typical densities in vacuum systems, in the upper atmosphere, in "outer space," etc.

2c. Which is the best definition of an ideal gas?
a) A gas whose atoms or molecules do not attract or repel one another.
b) A gas whose atoms (or molecules) are not attracted or repelled by the walls of any container.
c) A gas composed only of single atoms (no molecules).
d) A gas whose total energy does not depend on pressure.
e) all of the above
f) none of the above

Questions like this frequently require deliberately discussing each of the choices one by one. Tabulating the percentage of students who pick each of the choices can help direct the discussion into the most profitable areas. Most texts indicate that ideal gases have elastic collisions with the container walls so that momentum and kinetic energy are conserved, so discussion of choice (b) requires connecting the textbook description to the notion of molecular attraction to, or repulsion by, the walls. Students who connect the temperature T to the kinetic energy of the gas and then think about the ideal gas law may wrestle with choice (d) because the product PV is proportional to T. It is also important to bring in the fact that most gases (such as air) very nearly obey the ideal gas law over a wide range of pressures, so considering a gas to be "ideal" is not necessarily a case of unrealistic "non-real world" approximations.

3a. For an ideal gas, the change in internal energy is given by $\Delta U = nC_V\Delta T$ for **any** process. Please explain how this can be true for processes that are not isochoric.

Most students will need to go back to the textbook to try to answer this question. Many will need to refresh themselves on the meaning of "isochoric." Many will not realize that the keys to understanding this statement are the first law of thermodynamics and the definition of specific heat. It can be helpful to contrast an isochoric process, which involves no work and in which heat added only changes the internal energy, and an isobaric process, which does involve work and in which heat added both changes the internal energy and performs work. A good class activity is to graph isochoric and isobaric processes on a PV diagram and describe the changes in internal energy U, heat Q, and work W, corresponding to each process, and to write each in terms of (change in) temperature through use of the ideal gas law.

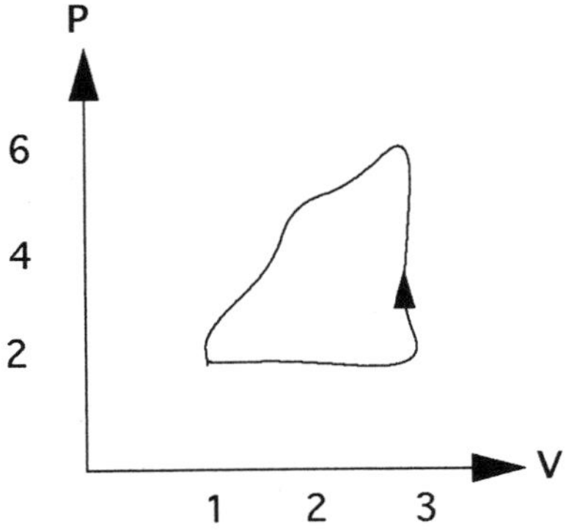

3b. Estimate the total work by a system taken around the cycle shown in the figure. Does the system have to be an ideal gas?

Students who work only with "canned" or memorized formulas will be unsure how to best approach this problem. Those who recognize the work as the area enclosed in the PV diagram will often be able to correctly approximate the cycle as a triangle and reasonably estimate the enclosed area. Some students will neglect the fact that the triangle has its lowest corner at (1,2) and will use (1/2)(3)(6) rather than (1/2)(3-1)(6-2) to estimate the area. Since most textbook examples deal only with ideal gases, many students will be unsure about the necessity of the "ideal" nature of the gas. An interesting extension of this is to ask how the temperature of the system might be changing during various parts of the cycle shown.

3c. If a process is closed (system returns to its original state), which of the following are zero?
a) The work done by the system.
b) The work done on the system.
c) The heat added to the system.
d) The change in energy of the system.
e) none of the above

Some students would like to believe that if a process is closed, then the work, the heat added or subtracted, and the change in energy of the system should all be zero. To be able to answer correctly, students must understand what "returning to its original state" means, and this is difficult for some. Students who understand that the system must return to its original temperature generally understand that the system also returns to its original energy and will therefore choose (d). Also, students who recognize that this question really deals with the first law of thermodynamics will generally choose (d). This question provides a good opportunity to spiral back to the notion of a conservative force and the fact that the work done by a conservative force acting over a closed process is zero. This provides a nice foundation on which to restate the first law of thermodynamics, which introduces the quantity of heat that wasn't considered in mechanics.

4a. Is melting an ice cube an irreversible process? How about freezing one?

Most students have a difficult time understanding reversible and irreversible processes. Since the concept of reversibility is generally new to the students, they will base almost their entire understanding on what they glean from the textbook. Many believe that "reversible" means "idealized," while "irreversible" implies "real world." Some textbooks discuss reversibility by describing making a movie of a process occurring. In such discussions, if one could run the movie backward and reasonably expect to see the in-reverse process actually occur, then that process is described as reversible. Of course, people actually see ice both melt and freeze, so a careful in-class discussion of reversible and irreversible processes will almost always be necessary in such cases. If, on the other hand, the textbook describes a reversible process as one in which both the system and its environment can be returned to their initial states, it will be somewhat easier for students to see that any process that occurs spontaneously, like the melting of ice, is irreversible. Here is a case when simple demonstrations, such as shaking layers of cinnamon and sugar together, are particularly helpful.

4b. On a hot summer day it is 90°F outside and 70°F inside. Estimate the electrical power consumed by an air conditioner that moves 5000 BTU/hr under these conditions.

Several aspects of this problem are difficult for students. The first is the need to consider an air conditioner as a "heat pump." Perhaps the most difficult task is to relate the quantities in the problem to the quantities used to describe heat pumps and refrigerators in the textbook. In particular, many students will confuse the rate at which heat is extracted from the cooler reservoir with the power input to the device and will confuse energies with rates of energy transfer. Then, many will be uncomfortable considering the air conditioner as a Carnot refrigerator (to calculate a lower bound on the power required) and will be uncertain how to deal with the "real" numbers given in the problem. Finally, unit conversions are a major obstacle. Many students will incorrectly assume that they can leave the temperatures in Fahrenheit because the coefficient that relates the input work to the heat extracted depends only on ratios of temperatures.

4c. The coefficient of performance of a refrigerator is a number between

a) zero and infinity.
b) one and infinity.
c) zero and one.
d) minus infinity and infinity.
e) none of the above

This is a good time to reemphasize that coefficients of performance (COPs) and efficiencies are simply defined as the ratio of (what you get out of the device) to (what you have to pay [input to the device] for that output). Since the COP is then just the ratio of the heat extracted at the lower temperature to the work put into the refrigerator, it is given by $T_l/(T_h-T_l)$. *Students who know this will generally pick (a) as the correct choice, but they will also sometimes choose (b). Students who confuse the COP with the engine efficiency will pick (c).*

5a. Is it possible to convert all the heat put into a system into work? If so, under what conditions? If not, why not? Please explain your answers.

Students who do not understand the significance of the second law of thermodynamics often believe that all heat should be able to be converted to work, because this would still satisfy conservation of energy. Some believe that a "perfect" (e.g., Carnot) engine would be able to do this, while "real" devices cannot because of things like friction that produce heat. This question provides good opportunities to delve into the difference between heat and work and to reinforce the concept of entropy.

5b. Suppose you have a standard refrigerator that consumes about 160 W in normal operation. Use reasonable estimates for T_h and T_l to estimate an upper limit for how much heat (Q_l, measured in Joules) could be extracted from the refrigerator per second.

The most frequent mistake students make with this problem is improper use of units. Many fall into the trap of converting their temperature estimates, which are generally quite reasonable when expressed in degrees Fahrenheit, into degrees Celsius, and then assuming that the needed temperature ratio will be correct. Subtler points include confusing work with power and energy with rates of energy transfer. Also, another point worthy of a bit of class time is the fact that the high temperature is actually somewhat warmer than typical room temperature, which is why the air at the back of the refrigerator feels warmer than the ambient air. One could imagine asking this same question with wording that makes fewer connections to the textbook "jargon"; this would certainly make the problem more difficult for most students.

5c. Which of the following is a true statement about Carnot efficiencies and coefficients?

a) Both the Carnot efficiency and the Carnot coefficient are always less than or equal to one.

b) The Carnot efficiency is always less than or equal to one, but the Carnot coefficient is always greater than or equal to one.

c) The Carnot efficiency is always less than or equal to one, but the Carnot coefficient can be greater than, less than, or equal to one.

d) The Carnot efficiency is always greater than or equal to one, but the Carnot coefficient can be greater than, less than, or equal to one.

Students often confuse the efficiency and the coefficient of performance (COP), and this question brings the confusion to the forefront. Discussion about what the efficiency is used to describe (an engine) and what the COP is used to describe (refrigerator, heat pump) is often fruitful. Sometimes I use this question as a springboard into a class group activity in which the students rederive each of the efficiency and COP expressions to reinforce their definitions. Extensions into the comparison between ideal and real efficiencies and COPs are natural.

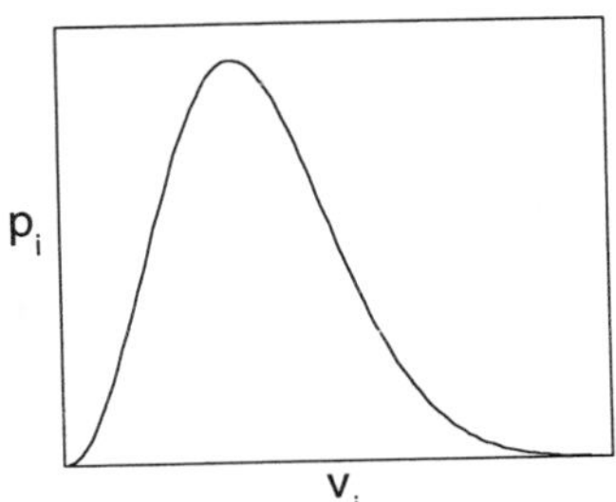

6a. The figure shows a typical Maxwell speed distribution. How would this distribution change for the same gas, but at higher temperature? Also, briefly explain where the most probable, average, and root-mean-square speeds would fall on this graph. Be sure to state their relative order.

Most students correctly indicate that the peak of the distribution shifts toward greater speeds (to the right) for higher temperatures. However, many fail to state that the peak also shifts downward (to a lower number of particles per unit speed interval) and that the entire distribution broadens as the temperature increases. A good classroom discussion generally ensues when one brings up the idea that the area under the curve must correspond to the total number of gas molecules and so must remain the same for distributions at different temperatures. Many textbooks include a figure that shows the relative positions of the most probable, average, and rms speeds, so most students will be able to correctly state their relative order. However, most students do not appreciate the significance of each of the quantities, so this question provides a convenient springboard into a discussion or activity addressing why each speed is defined and what it means. This is also a good time to address the fact that the distribution is not symmetric, because of the high speed "tail," and the implications of this for seemingly unrelated situations like evaporation and gas molecules escaping from a planet's atmosphere.

6b. Estimate the speed of a hydrogen molecule at room temperature. At this speed, if the hydrogen molecule didn't run into any other molecules, about how long would it take it to get from one end of your squadron room to the other?

Some students will be uncertain which speed—most probable, average, or root-mean-square—to estimate. Typical mistakes include using the mass of a hydrogen atom rather than a hydrogen molecule, forgetting to express room temperature in Kelvin, and forgetting to take the square root. Students who complete the calculation correctly are usually surprised by the shortness of the transit time across the room. This provides a natural opportunity to talk about gas molecule collisions, random walk motion, the speed with which smells travel across a room, etc.

6c. The temperature of an ideal gas is proportional to
a) the average velocity of the atoms.
b) the average momentum of the atoms.
c) the average energy of the atoms.
d) none of the above

Students who have a vague concept that higher temperature corresponds to gas molecules moving faster may be tempted to choose either (a) or (b). Those who have a more complete understanding of kinetic theory will usually recognize that temperature is proportional to the average kinetic energy of the gas atoms or molecules. The "loose" wording in choice (c), which indicates only "energy" as opposed to "kinetic energy," leaves open some ambiguity in the question. This provides a good opportunity to emphasize the importance of clarity and completeness of expression. (Note: This is the type of question for which students might be encouraged to use the general comments box at the end of the WarmUp to "defend" their choice. If they realize the ambiguity, they can clarify their interpretation of the choices and explain why they answered as they did.)

Electricity and Magnetism

1a. Can there be an electric field at a point where there is no charge? Can there be a charge at a place where there is no field? Please write a one- or two-sentence answer to each of these questions.

Many students believe that the field does not exist if there is no test charge to feel a force. Most believe that "self fields" must be counted and thus claim that no charge can experience zero field. Rooting out problems such as these at the earliest possible time is among the greatest advantages of using WarmUps.

1b. Let's say you are holding two tennis balls (one in each hand), and let's say these balls each have a charge Q. Estimate the maximum value of Q so that the balls do not repel each other so hard that you can't hold on to them.

Many students will guess wildly at the force they can withstand, or they will forget to convert to SI units. Even students who get this wrong appreciate the opportunity to get "personal" with the abstract notions of electric field and force. A useful extension is, "If you are holding a ball with a 1 mC charge, how large a field can you withstand?"

1c. An insulator is a material that ...
a) cannot feel an electrical force.
b) is not penetrated by electric fields.
c) cannot carry an electric charge.
d) a, b, and c are all correct
e) none of the above

Most students believe that insulators "block" all things electrical and will pick a or b. Many also are unfamiliar with the use of "carry" as a synonym for hold, and will pick c or d.

2a. The flow of water through a pipe is measured in cubic meters per second, and it is given by Flow = vA where v is the velocity and A is the cross-sectional area of the pipe. Please explain the idea of electric flux by making an analogy to the flow of water.

Until they have been asked to discuss good answers to this type of question, most students will give very vague answers like, "The electric flux is just like the flow because it also depends on the area of something." I use questions of this type at several times during the semester and encourage students to develop answers that use physical arguments, draw specific analogies, etc.

2b. Please estimate the volume and surface area of a basketball.

Many students simply do not know that $V = 4\pi r^3/3$ and $A = 4\pi r^2$. Those that do know often estimate the diameter and forget to divide by 2. It is best to get these details out in the open before Gauss's law becomes the issue.

2c. Which of the following is true of a perfect conductor?
a) There cannot be an electric field inside.
b) There cannot be any electric charges inside.
c) There cannot be an excess electric charge inside.
d) There can be no electric charge on the surface.
e) Two of the above choices are correct.

Many students fail to distinguish among charge, excess charge, and free charge. We use this question to open a discussion of the differences. We also emphasize the need for free charges to account for conductivity and the absence of fields in a conductor.

3a. The electrical potential energy stored in a capacitor is given by $U = (1/2)CV^2$. It is also given by $U = (1/2C)Q^2$. Is the energy really proportional to C or to $1/C$, or is neither really true?

Many students would like to believe that one of these is inherently superior. This is a good opportunity to emphasize that memorizing formulas is a poor method of studying physics. It is worth coming back to this while discussing power in conductors and asking the same question again: $P = I^2R$ vs. $P = V^2/R$.

3b. You and a close friend stand facing each other. You are as close as you can get without actually touching. If a wire is attached to each of you, you can act as the two conductors in a capacitor. Estimate the capacitance of this "human capacitor."

Most students hesitate to make the approximations necessary to do this. Many will not see the possibility of treating the pair as parallel plates. Some will object that people are not conductors.

3c. What is the relationship between the charges on the electrodes of a cylindrical capacitor?

a) Larger on the outer cylinder.
b) Larger on the inner cylinder.
c) Exactly the same.
d) Equal in magnitude but opposite in sign.
e) Any relationship is possible.

Many students would like to believe that the charge densities must be equal rather than the total charges. They will pick a. Others will answer c rather than d. Answer e is very appealing to students who are feeling overwhelmed. Discussing this answer is a good opportunity to stress the certainty possible when laws are correctly applied.

4a. According to the book, current is given by the change in electric charge over the change in time, i.e., $I = \Delta Q/\Delta t$. Does this mean that the amount of charge in a wire or a circuit changes with time? If not, what does it mean?

This definition (and its equivalent in differentials) often confuses students, even those who have a reasonable feel for current. This question is a good way to open a discussion of the microscopic model of current, the difference between velocity and drift velocity, etc.

4b. Estimate the resistance of a typical lightbulb.

Most students try to make this much more complicated than it is. They try to estimate the geometry and resistivity of the filament rather than simply working with common values of voltage and power. Students who take this route may be praised for creating an opportunity to discuss the dependence of resistivity on temperature.

4c. When current passes through a resistor, it

a) increases.
b) decreases.
c) stays the same.
d) any of the above depending on the circuit

Many students believe that current is "used up" as it passes through a resistor and will pick b. As in #3c, students that feel they are lost are likely to pick d rather than admit to having any confidence about anything. Discussing this problem is also a good time to emphasize that current passes through a battery as well. Many students believe that batteries are like charged capacitors, with a reservoir of positive charges in one place and a corresponding reservoir of negative charges elsewhere.

5a. Two resistors of very different value are connected in parallel. Will the resistance of the pair be closer to the value of the larger resistor or the smaller one. Why?

Many students will add "sample values" and determine that the small value dominates but provide no physical interpretation. Discussing this in class is a valuable exercise in seeking understanding in general. Some students will make the mistake of guessing that large values always win.

5b. A 1 mF capacitor is charged up to 1000 V and then allowed to discharge through a 100 kΩ resistor. Estimate how long it takes before the voltage across the capacitor has dropped to a safe value. (Note: This is why you should *not* poke around inside a TV set if you do not know what you are doing.)

Many students will calculate $\tau = RC$ correctly but will not think about what might be safe. This is another question that students appreciate when they see the connection between a numbered equation and a personal experience.

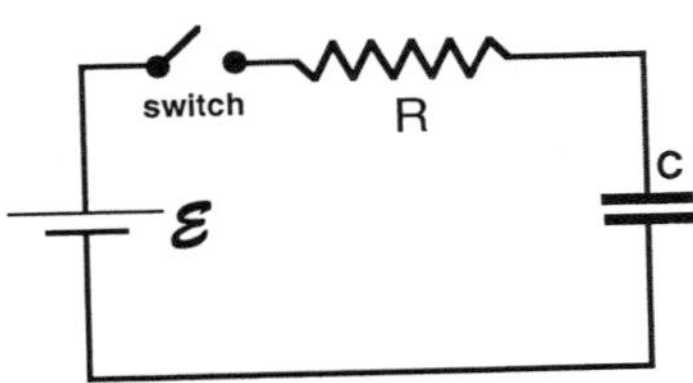

5c. Consider a circuit like the one shown. A long time after the switch is thrown, the capacitor will have an (approximately) constant charge (Q) and voltage (V). If the value of the resistor is increased, which of the following statements is true?

a) Q will be less, but V will be unaffected.
b) V will be less, but Q will be unaffected.
c) Q and V will both be less.
d) Neither Q nor V will be affected at all.
e) V will be less, but Q will be greater.
f) Q will be less, but V will be greater.
g) Both Q and V will be greater.

Many students believe that one or both of these quantities will decrease due to increasing R. This is a good time to emphasize the process of testing limiting values of solutions.

6a. The force F experienced by a particle with charge q moving with velocity v in a magnetic field B is given by the equation $F = qv \times B$. Of course, the force is zero if the particle is not moving ($v = 0$) or if there is no magnetic field ($B = 0$). There are other cases where a particle will not experience any force even when moving in a magnetic field. Please describe these situations.

Many students will get $q = 0$, some will mention "v in the direction of B," and some will say perpendicular! Few will explicitly suggest v antiparallel to B. We usually work directly from this question to the conclusion that the B field does not do work.

6b. The electron gun in a color TV accelerates the electrons through about 12,000 V. Estimate the acceleration of such an electron due to the magnetic field of the Earth.

Many students will not look up the field of the Earth if it has not been made obvious in lecture or in your text. Many will also have trouble with the notion of picking an arbitrary angle. Some will have ingenious reasons for deciding this should be zero. There are several good extensions of this problem. One can ask about the time of flight and net deflection between gun and screen. This is about 5 mm, so TVs and monitors have to compensate internally. How this is accomplished is a good research question.

6c. A proton moving downward enters a region of space with magnetic field B that points eastward. In which direction is the force on the proton?
a) The force is zero.
b) The force is upward.
c) The force is to the north.
d) The force is to the south.

Many students will lose their way with the right-hand rule before they determine that the correct answer is south. The extra negative sign introduced by using an electron or other negatively charged particle causes even more confusion.

7a. The text shows the magnetic field lines around a long, straight wire as closed circles rather than individual vectors. Could you redraw this figure showing vectors that each have a beginning and an end? Describe the picture as best you can.

Many students will get the tangential direction but fail to indicate the 1/r dependence with decreasing lengths.

7b. Estimate the force of the Earth's magnetic field on a 10-cm segment of a typical wire in your home.

One amp gives a few micronewtons, but students will estimate currents from a milliamp to hundreds of amps. The issue of picking an angle is the same as in #5c.

Asking about the direction of the force (even raising the issue of ac current) makes a nice extension.

7c. If a flat loop of wire carrying a uniform current feels a net force due to a magnetic field, which of the following must be true?

a) The magnetic field is perpendicular to the plane of the loop.
b) The magnetic field is in the plane of the loop.
c) The magnetic field is constant over the loop.
d) The magnetic field is not constant over the loop.
e) None of the above conditions are really necessary.
f) It is impossible for a loop to feel a net force, no matter what.

Many students will correctly pick d but not really see what the possibilities are. Asking students to sketch in the forces on a field in a gradient is worthwhile, and changing the geometry is a good extension.

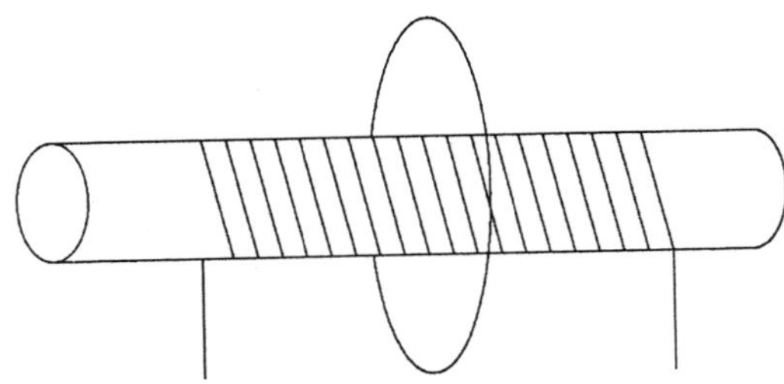

8a. A wire loop surrounds a solenoid wound around a wooden bar. If the current in the solenoid (I) increases, will there be a current in the loop? Which way?

Many students will think that no current will flow, saying that the magnetic field is confined by the solenoid and hence does not touch the loop. Of course, the field is not confined in this short solenoid, but even if it were, students must be reminded that it is flux, not field, that causes induction. Hence, a discussion of the infinite solenoid case is worthwhile. Beyond this, many students will simply have a hard time describing the direction of the induced current. We encourage "up across the front" or counterclockwise viewed from the left.

8b. Estimate the magnetic flux (due to the Earth's field) through your body (back to front) if you stand facing north. What will be the net change in flux if you turn and face south?

If the student views the body as a 3-D object, then the net flux is zero in either case. If the student considers a silhouette, then the flux can be estimated from the Earth's field and the area of the silhouette, with the net change being 2φ (many students will get the initial estimate but put zero here). This is a good opportunity to talk about communications in addition to talking about flux.

8c. A uniform magnetic field can produce an EMF in a conducting loop if
a) the field changes in magnitude.
b) the loop changes position within the field.
c) the loop changes size.
d) the loop is removed from the field.
e) all of the above
f) all except b
g) all except c
h) all except d

There are two common misconceptions that can be addressed in a discussion of this problem. First, many students make the mistake of thinking that any magnetic field always induces a current if there is a flux; they confuse the change in flux with the flux itself. The second common problem causes students to choose answer e rather than f. They feel that the flux must be changing if it is coming from "a different part of the field."

9a. Here's one way of understanding a capacitor: It is a device that won't let the voltage between two points change too rapidly, because it stores up charge and has $V = Q/C$. The charge cannot be changed instantaneously, so the voltage cannot either. Try to describe an inductor in a similar way—that is, say what cannot be changed rapidly and why.

If this is the third or fourth such question in the semester, students should be able to make specific analogies, e.g., "An inductor would be a device that won't let the current change rapidly, but in this case it is magnetic flux that cannot be changed too instantaneously, like the charge in the example." This is a good time to actually sketch the lines of induced E field and show how they tend to maintain the current.

9b. Estimate the inductance of a solenoid made by buying a typical spool of wire from a hardware store and winding it carefully around a broom handle.

This is a good opportunity to discuss the dependence of L on winding density rather than number of turns.

9c. A resistor, an inductor, an ideal battery, and a switch are connected in series. Just after the switch is closed, which of the following are zero?
a) the current and the voltage across the inductor
b) the current and the voltage across the resistor
c) the voltage across the inductor and the voltage across the resistor
d) all of the above

*Many students will have the general notion that the inductor "keeps anything from happening" and choose a, c, or d. This question can lead into a discussion of what the voltage **has** to be across the inductor to suppress current changes. It is* *also a good time to remind students that Ohm's law requires V to be proportional to I only for resistors.*

10a. In a wave on a string, the amplitude is the maximum displacement of a point on the string from its equilibrium position. What is the amplitude of an electromagnetic wave?

Many students will write down an acceptable answer (E, B, both A and B) without really understanding what a plane wave (for instance) is. The pictures in most texts confuse students. They see the sine function as the shape of the E field, or the E field on the axis, without realizing that the sine function, in this case, applies to whole parallel planes.

10b. Estimate the wavelength of your favorite FM radio station and your favorite AM station.

Most students can do this, provided they realize that the numbers on the dial refer to MHz and kHz. This is a good time to have a qualitative discussion of things like antenna sizes, reception under bridges, etc.

10c. The choices below list several different kinds of electromagnetic radiation. Which list is in order of increasing wavelength?

a) gamma rays, infrared, ultraviolet, visible light
b) radio waves, microwave, visible light, X rays
c) X rays, ultraviolet, red visible light, green visible light
d) microwaves, ultraviolet, X rays, gamma rays
e) none of the above

Many students, even those who did 10b correctly, will fall for the reverse order trap set in choices b and d. Others will choose c, missing the reversal of red and green. We have had a few students complain: "I thought none of the above was never the right answer."

Optics

1a. Explain how we (in physics) define the angle between a beam of light and a surface that it hits.

The student answer we expect goes something like this: "We determine the point where the ray will hit the surface. Next we construct a line that is perpendicular to the surface at that point. This line is called the normal to the surface at the point in question. The incident direction of light is determined by the angle between the ray and the normal." This easy WarmUp question is designed to ensure that students read the chapter carefully and describe, in their own words, what the terms "ray of light," "optical medium," and "normal to the surface" mean to them. In our experience, students can learn to manipulate ray diagrams and to an-

swer questions about them without much of an understanding about the relation between ray diagrams and real-world behavior of light. Having students verbalize the abstract relations and discuss their responses in class starts the process of building this relation.

1b. Estimate the time it takes for light to do one lap in a swimming pool (underwater).

An easy estimate exercise ensuring that students notice in the reading that the speed of light depends on the medium. To make the estimate, they have to calculate speed of light in water from the index of refraction. The answer is: about 0.2 microsecond.

1c. Total internal reflection can occur when light traveling in a medium reaches a boundary
a) with a medium having a lower index of refraction.
b) with a medium having a higher index of refraction.
c) with any other medium.
d) that is smooth enough.
e) none of the above

This is a simple question that students have to wrestle with before any classroom discussion of geometrical optics occurs. There are students that will have problems with this. Struggling with this question will make them more receptive to the classroom discussion of Snell's Law.

2a. Please explain in your own words what a focal length is. Try not to use any equations or refer to a specific type of mirror or lens.

This is another "What does this jargon mean to you" question. A student that approaches this question conscientiously has to come to terms (at least in first approximation) with the notions of parallel rays, plane waves, far away light sources, optical axes, converging and diverging devices (both reflecting and refracting), and real and virtual points. We do not expect perfect answers here.

2b. Estimate the focal length of a typical "magnifying mirror" that you might purchase from a drugstore.

Students may or may not have seen a cosmetic mirror. Many of them have. Attempts at this estimate will include anything from "empirical" observation (e.g. Where does the image blur?) to the use of the mirror equation. 20 or 30 cm is a good estimate.

2c. The focal length of a plane mirror is
a) undefined.
b) zero.
c) one.
d) infinity.
e) none of the above

This may be a hard one. A plane mirror does not focus. Yet it is a limiting case of a curved mirror as the radius of curvature gets very large. Consequently, it is the limiting case of a curved mirror as the focal length gets very large. So the answer must be d. Does the mirror equation still make sense? A plane mirror forms a virtual image with image distance having the same magnitude as the object distance. What does that make f?

3a. If we use a lens with a focal length of 25 cm to form an image of an object that is 5 cm tall, how far away must the object be for the object distance to be approximately infinity? Does your answer depend on the size of the object, the focal length, or both?

This one is also hard. Nothing in real life is infinite. How do we decide what is approximately infinite? If the object distance were truly infinite, the image would form exactly at the focal point (for a point object situated on the optical axis.) The object distance will be approximately infinite if the image distance is approximately 25 cm. What that is depends on how accurately we determine distances. How about a real image distance of 25.0001 cm? Infinity is 6 million cm away.

3b. The lens at the front of your eye must produce a sharp image on your retina at the back of your eyeball. If your vision is good, it can do this whether the object is 1 ft or 100 yd away. Estimate the range of focal lengths that a "good" eye can adopt.

Do this in inches. The size of the eyeball is about 1 in. To image objects very far away, the focal length of the eye has to be about an inch = 2.5 cm = 0.025 m (in terms of power, about 40 diopters). To image an object 12 in. away, the power has to be about 43 diopters.

3c. A flat piece of glass has a focal length that is

a) undefined.
b) zero.
c) one.
d) infinity.
e) none of the above

We can reason as we did in question 2c. Again, the answer is d.

Chapter 9: Puzzles

General Principles

Most of the principles governing the design of a WarmUp apply equally well to the design of a puzzle. Good puzzles often (but not exclusively) involve a real-world scenario, and they should be good discussion starters. The primary differences between the two are in the complexity of the question and the precision expected in the answers. A WarmUp is intended to introduce a subject, while a puzzle should end it. Thus, puzzles should be more subtle and should require integration of many ideas. In turn, puzzles should be evaluated more rigorously. The answers should be technically correct and well stated. In our classes at IUPUI, about 80% of the students get credit for each WarmUp; only about 25% get credit for each puzzle.

Mechanics

1. The standard geographical coordinates of Chicago are:
Latitude: 41 degrees 50 minutes
Longitude: 87 degrees 45 minutes

What are the x, y, z coordinates of Chicago in a coordinate system centered at the center of the Earth with the z-axis pointing from the South Pole to the North Pole and the x-axis passing through the zero longitude meridian pointing away from the Earth, into space. Please answer this in words, not equations, briefly explaining how you obtained your answer.

We use this puzzle to close the discussion of frames of reference and vector algebra. The students are expected to describe the transformation from a spherical coordinate system to a Cartesian system. In class we present a sampling of student responses and then work out the problem in detail. The process includes a review of trigonometry and the newly learned vector algebra. Our students find this puzzle moderately difficult.

2. An airplane flying at constant air speed from Indianapolis to St. Louis in calm weather (no wind of any kind) would log the same flying time for both legs of the trip. Suppose the same trip is taken when there is wind from the west. How would the total (round-trip) time in windy weather compare with the total time in calm weather? Please answer this in words, not equations, briefly explaining how you obtained your answer.

This puzzle closes the first week of kinematics. A fair number of students argue that the wind effects will cancel out on a round-trip. This is a naïve symmetry argument that has to be analyzed and disposed of. Very few students in our classes think of setting up the algebraic solution for this question. Even fewer check their reasoning by examining extreme cases. What happens if the plane air speed is slower than the wind speed? If they have an algebraic solution, does it reduce to the correct answer if the wind speed is zero?

3. You probably know by now, from reading the book and from working the projectile motion problems, that the maximum range for a projectile is achieved when the projectile is fired at 45 degrees. This is true if the launch starts and ends at the same altitude. What if the target is at a lower elevation? Is the optimal angle still 45 degrees? If not, is it more? Less?

This puzzle is discussed at length in Chapter 6. It is presented to the students at the end of 2D kinematics. In our experience, students find this puzzle difficult. Notice that we do not ask for the optimum angle. To answer that question, we would need to know the two elevations. However, knowing that the target elevation is below the launch elevation, we can show that the optimum angle must be less than 45 degrees. By the time this puzzle comes up, we are three to four weeks into the semester. Hopefully we have come a long way developing our problem-solving skills. This is a good place to put some of those tools to work.

4. A 10-kg mass and a 5-kg mass are suspended from pulleys as shown in the diagram. A small metric scale (calibrated in newtons) is spliced into the rope joining the two masses. Assume that the mass of the scale is negligibe. What does the scale read after the masses are released from rest?

This is a fairly easy puzzle. The student answers are full of surprises, reminiscent of the Force Concept Inventory results. We give this puzzle somewhere near the middle of the discussion of Newton's laws. It is basically a paraphrase of the Atwood's machine question. Try it.

5. Bill is riding in a railroad car, throwing a ball against the car wall. The train is moving at 20 m/s to the right. Bill is throwing the ball at 8 m/s to the left. According to Bill, his 0.145-kg baseball has 4.64 J of kinetic energy. His brother is standing on the ground disagreeing. According to him, the baseball has 10.44 J of kinetic energy. Who is right?

Again, an easy question with a surprising outcome. Both boys are right. But about half of our students miss this one. We use this question to examine the definitions of physical quantities more carefully, to discuss how they might depend on the frame of reference. (Do vector quantities depend on the reference frame? In what sense?)

6. You are standing on a log, and a friend is trying to knock you off. He throws the ball at you. You can catch it, or you can let it bounce off of you. Which is more likely to topple you—catching the ball or letting it bounce off?

This is yet another superficially easy question that many students find puzzling. Most of our students first guess that you are more likely to get knocked off if you catch the ball. They justify their answers by arguing that whatever it is that the ball brings you, you don't want to absorb it. This is a good place to talk about the intuitive meaning of momentum and energy. What do shock absorbers absorb? The notion of shock is related to the notion of force. How is force related to momentum and to kinetic energy? What does the ball do to you to topple you?

Since you will suffer a change of momentum equal and opposite to the change of momentum experienced by the ball, it is to your advantage to minimize that change. Bouncing the ball requires twice as much change as catching it.

7. Hold a basketball in one hand, chest high. Hold a baseball in the other hand about two inches above the basketball. Drop them simultaneously onto a hard floor. The basketball will rebound and collide with the baseball above it. How fast will the baseball rebound? Assume that the basketball is three to four times heavier than the baseball.
The result will surprise you. Don't do this in the house!

This is a familiar exercise that has been around for years. This is a difficult puzzle. Students will approach it in a variety of ways. Very few students will deal with this algebraically, comparing the speed of the baseball to the speed of the basketball as it touches the floor, which is about 5 m/s. Most of our students attempt to solve this piecemeal and usually bog down in the algebra. Devoting the entire hour to this problem gives an opportunity for a thorough review of just about everything covered in the course so far. It also affords a chance to review the problem-solving techniques we have practiced.

8. You are given two identical-looking metal cylinders and a long rope. The cylinders have the same size and shape, and they weigh the same. You are told that one of them is hollow, the other is solid. How would you determine which is which, using only the rope and the two cylinders?

This is an easy puzzle with a short answer. Wrap the rope around both cylinders and cause them to rotate. All else being equal, the hollow one will accelerate more slowly because it has a greater moment of inertia. We extend this puzzle to include assorted scenarios involving rolling without slipping.

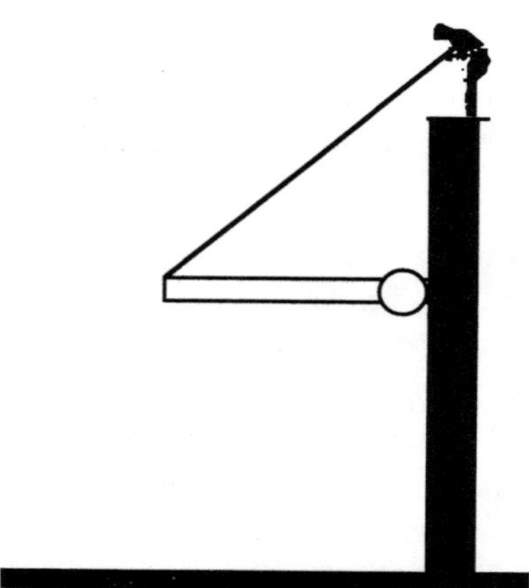

9. Is it possible to balance a horizontal bar against a smooth vertical wall with one cable, attached to the far end of the bar? The force that a smooth vertical wall

exerts on the bar cannot have a vertical component (the bar is balanced if there is no net force on the bar and if there is no net torque about *any* axis).

Most students' intuition tells them that the bar cannot be so balanced. Some even try it. However, a large number of students in our classes believe that without the presence of a vertical force at the wall, the bar cannot be balanced in any configuration. This is false. A horizontal force at the wall can fix the torque requirement if the bar is tilted. Which way?

10. Consider a satellite in a circular orbit about the Earth. If NASA wants to move the satellite into a higher orbit, they have to boost the satellite speed. Yet when the satellite is inserted into the new (higher) orbit, its speed is actually less then it was in the old (lower) orbit. Is this correct? If the answer is yes, can you explain how the satellite slows down.

This puzzle is a good introduction to the lesson that summarizes orbital mechanics at the introductory level. In the higher orbit, the kinetic energy of the satellite is less than it was in the lower orbit, but the potential energy is higher. The "missing" kinetic energy was used up in work done against gravity. Yet another visit paid the work-energy theorem. Our students work this puzzle in conjunction with a lab exercise dealing with the Hohmann transfer.

Thermodynamics

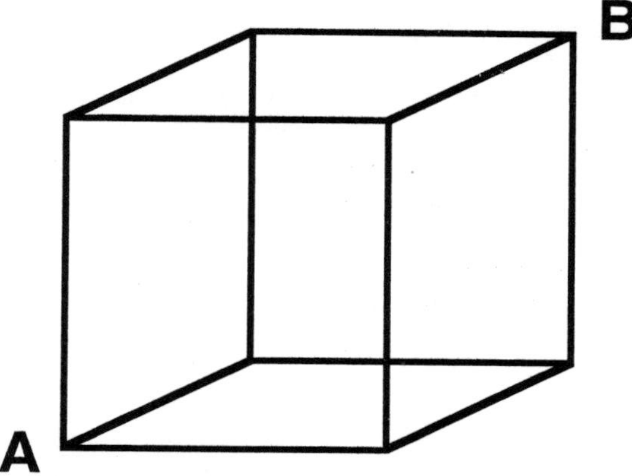

1. Twelve identical aluminum bars are welded together in the form of a cube. The bars have length 50 cm and radius 1 cm. If one corner (labeled A) is held at 100°C, and the opposite corner (labeled B) is held at 0°C, what will be the total heat flow from corner A to corner B?

Since thermodynamics comes after electricity and magnetism, this puzzle provides a nice opportunity to spiral back to the resistor cube and explore the similarities and differences between that system and the one in this puzzle. Drawing the analogies between the flow of current and the flow of heat can help the students improve their understanding of both concepts.

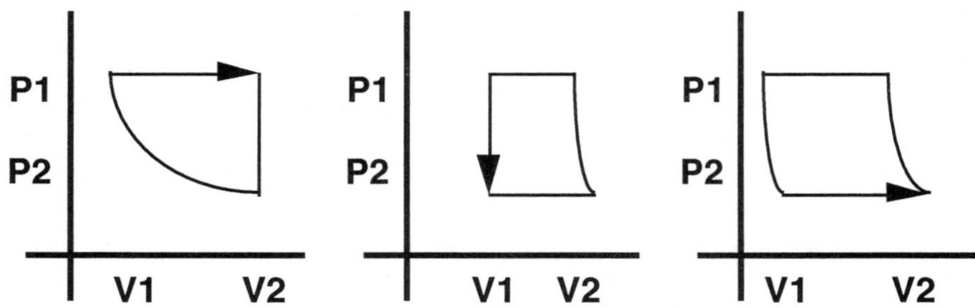

2. Three laboratory rats are placed inside three identical environmental chambers and subjected to three different cycles of changing volume and pressure. In each case, the limits on volume and pressure are the same, and in each case, the chamber returns to its original volume and pressure at the end of the cycle. PV diagrams describing the three cycles are shown in the figure. Which rat was subjected to the highest temperature, and which rat was subjected to the lowest temperature during the experiment? Which rat felt its ears "pop" the most?

We have found that our students require a great deal of practice with PV diagrams before they feel comfortable interpreting them and understanding what the various curves represent. In this puzzle, for example, most students do not immediately realize that points close to the origin represent low temperatures while points far from the origin represent greater temperatures. Asking the students to relate "ear popping" to something represented by a symbol and to then interpret the graph requires them to carry out a two-step process, each step of which can be challenging.

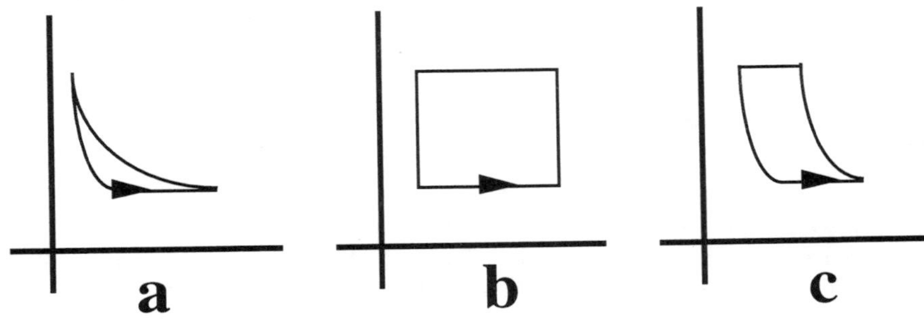

3. In honor of the 60th anniversary of the famous "War of the Worlds" radio broadcast in which the Earth was purportedly under massive attack by the Martians, suppose you have been abducted by Martians. The Martians wish to study human beings' tolerance to various environmental conditions. They place you in a sealed chamber that is subject to a series of changes in temperature, pressure, and volume. You can see the walls move, feel your ears pop, etc. Three possible PV diagrams for the air in your "cell" are shown above. The three marked points on each diagram correspond to identical conditions.

3a: Which of these processes involves the highest temperature? Which involves the lowest?
3b: Which of these processes would require the Martians to have a cooling system?
3c: Which of these processes would cost the Martians the most to complete? The least?

Both understanding the underlying physics and interpreting the wording in this puzzle make it challenging for many students. If students interpret the "cooling system" to be something that actually reduces the temperature in the cell over time, they will come up with different answers than if they interpret a cooling system to be something that extracts heat from the cell. The notion of "cost" is vague and requires clarification in the student answers. Discussing several student answers to this puzzle and focusing on the quality of their explanations of their assumptions can do a lot toward improving the students' technical communication skills.

4. One whimsical statement of the laws of thermodynamics is, "You can't win, you can't break even, and you can't get out of the game." Justify this statement in a clearly written paragraph.

A moderately difficult puzzle. This puzzle forces the students to grapple with the real meaning of the laws of thermodynamics rather than just to regurgitate or rephrase textbook definitions and statements. Perhaps the most difficult task is to determine which part of the quote refers to which law. Making the connection between everyday ideas ("break even," "win," etc.) and physics concepts ("energy," "conversion of energy to work," "thermodynamic equilibrium," etc.) is challenging but provides students with an opportunity to probe their own understanding. Discussion of student responses in class can be both entertaining and insight-provoking.

Electricity and Magnetism

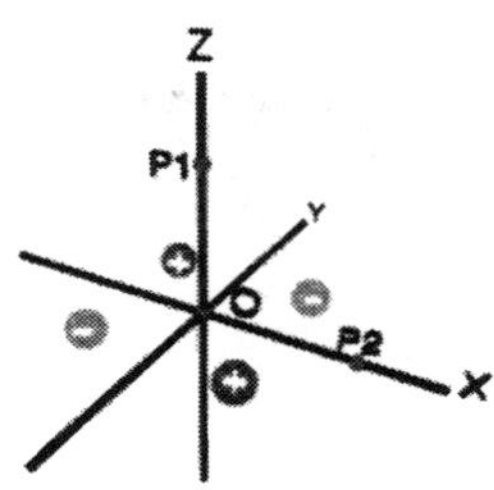

1. In the figure above, point charges are located at (1,–1,0) and (–1,1,0), and each carry a charge, Q. Other point charges are located at (1,1,0) and (–1,–1,0). They each carry a charge –Q. Consider the electric fields at three points that have no charges: O is the origin, P1 is (0,0,2), and P2 is (2,0,0). I would like to rank the points O, P1, and P2 in order of decreasing electric field strength. The correct answer is:

a)P1 > P2 > O	b)O > P1 > P2	c)P1 > O > P2
d)O > P2 > P1	e)O > P2 = P1	f)P2 > P1 = O
g)P2 > P1 > O	h)P1 = P2 > O	g)P1 = P2 = O

A moderately difficult puzzle. Many students will make one or more mistakes in analyzing this question. Some will ignore the vector nature of fields and identify the origin as the point with the greatest field strength, arguing that it is "closest to the most charges." These students will choose either b or d. Some will not recognize the symmetry at P1 and choose a or h. Others will see only the symmetry, and believe that the field must be zero at all points, choice g. Even students that determine the correct answer are surprised when they realize that the field at P2 is in the y direction. In discussing this question, we often begin with the origin, discuss the vector addition, then go on to consider a point just above the origin on the z-axis. Next, we consider points along the x-axis. In a talented class, this problem can open the door to discussions of multipoles.

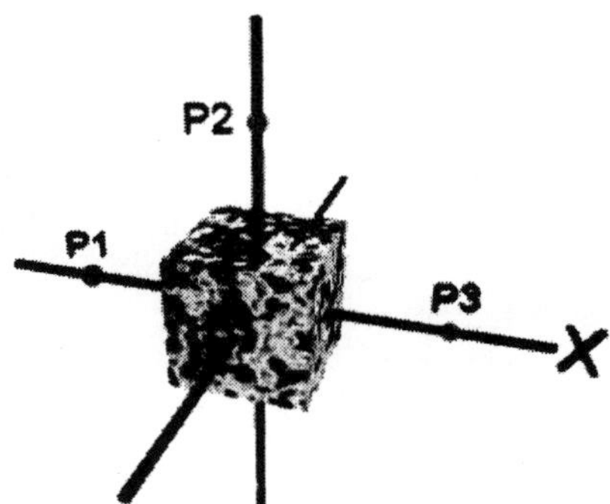

2. The figure above shows three points—P1, P2, and P3—in the vicinity of a camouflaged box. Measurements indicate that the electric field is in the –x direction at P1 and P3, but it is in the +x direction at P2. Please describe what charges might be in the box.

A relatively easy puzzle. The physicist's answer is a dipole at the origin, directed along the negative x-axis, but most students will describe less ideal charge distributions. We give credit for any answer that unambiguously produces the stated fields, e.g., "A positive charge just inside the +x side and an equal negative charge just inside the –x side." However, in the class discussion, we make an effort to praise the more elegant solution. We then use this problem as a launching point to discuss field lines and equipotentials, multipoles, etc.

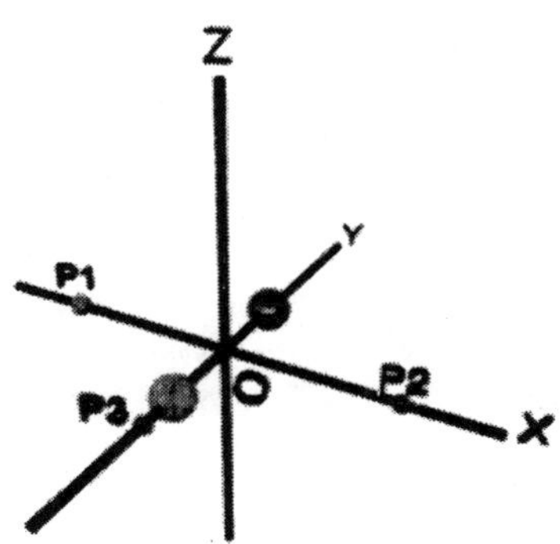

3. In the figure above, a ball carrying a charge Q sits at (0, –1) and a charge $-Q$ sits at (0, 1). We wish to move a small positive test charge q from point P1 (–2, 0) to point P2 (2, 0). There are two possible paths. Path **1** goes directly along the x-axis, and path **2** follows a triangular route from P1 to P3 (0, –1.5) to P2.

Which of the following is correct?
a) Path **1** requires less work.
b) Path **2** requires less work.
c) Paths **1** and **2** require an equal, positive amount of work.
d) Paths **1** and **2** require an equal, negative amount of work.
e) Paths **1** and **2** each require zero work.
f) There is not enough information to tell.

A relatively difficult puzzle. Many students find the notion of path-independence confusing and do not realize that the symmetrical placement of the endpoints assures zero work for all paths. Instead, most students focus on the forces, forget the dot product in the definition of work, and conclude that both paths require positive work, with path 2 more positive because of the large forces near P3. This puzzle offers a good opportunity to discuss potentials, review path-independence, and remind students of the crucial role played by the dot product in the definition of work. Good extensions are the work in moving from P3 to a symmetrically located point on the +y-axis, the same again for a negative test charge, and motions to or from infinity.

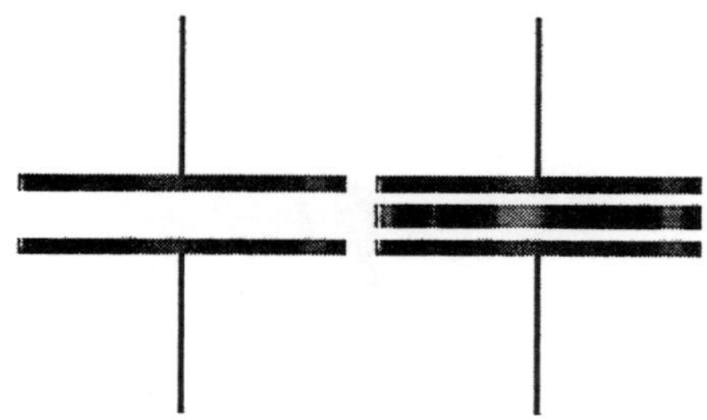

4. Consider the figures above. On the left is a plain parallel plate capacitor (area = A, gap = d). On the right is a system consisting of the same capacitor with a metal plate of thickness $d/2$ inserted into the center of the gap. How will the new capacitance compare with the old?

A moderately difficult puzzle. Many students will guess that C will be halved. Others will guess that C will double but fail to provide a correct argument. We find it valuable to discuss this problem by displaying two correct but very distinct student answers: one that analyzes this as two capacitors with gap d/4 in series, and another that uses Gauss's law to determine the charge distributions on the inner plate and calculates V from a presumed Q. A nice extension is the same problem with the plate inserted closer to one electrode than the other, or of variable thickness. A good question to add is, "What if that plate touches one of the electrodes?"

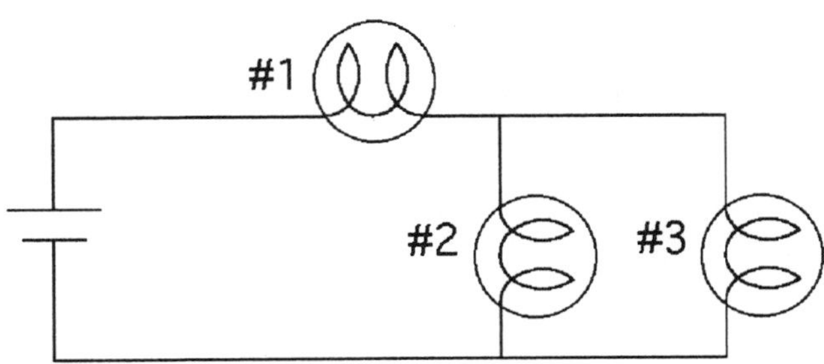

5. The circuit above shows three identical lightbulbs attached to an ideal battery. If bulb #2 burns out, which of the following will occur?

a) Bulbs 1 and 3 are unaffected. The total light emitted by the circuit decreases.
b) Bulbs 1 and 3 get brighter. The total light emitted by the circuit is unchanged.
c) Bulbs 1 and 3 both get dimmer. The total light emitted by the circuit decreases.
d) Bulb 1 gets dimmer, but bulb 3 gets brighter. The total light emitted by the circuit is unchanged.
e) Bulb 1 gets brighter, but bulb 3 gets dimmer. The total light emitted by the circuit is unchanged.
f) Bulb 1 gets dimmer, but bulb 3 gets brighter. The total light emitted by the circuit decreases.
g) Bulb 1 gets brighter, but bulb 3 gets dimmer. The total light emitted by the circuit decreases.
h) Bulb 1 is unaffected, but bulb 3 gets brighter. The total light emitted by the circuit increases.
i) none of the above

A moderately difficult puzzle. Eric Mazur describes a similar problem in his book Peer Instruction *[Mazur, 1996], and the results are similar as well. We find that many students conclude that bulb 3 gets brighter, but they do not predict the decrease in bulb 1. Even if they predict this, they suspect that the two effects cancel, leaving the total light output unchanged. Students who simply slog through and do all the calculations often get this problem right [choice (f)] and then express surprise at the result. In discussing this problem, it is worthwhile to help students focus on the total output without doing the calculations. That is, the net resistance of* *the circuit must increase, so the power for fixed V must decrease. This is a good opportunity to remind the class of the WarmUp question asking for a comparison of V^2/R vs. I^2R. (see WarmUp **3a** and discussion). A natural extension is adding a fourth bulb instead of subtracting one.*

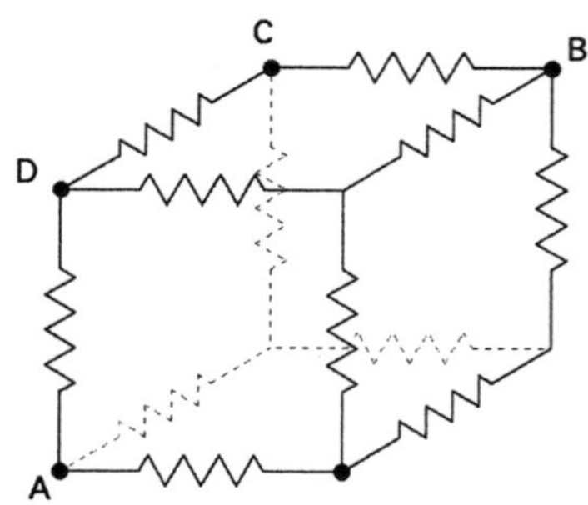

6. The figure above shows twelve identical resistors of value R attached to form a cube. Find the equivalent resistance of this network as measured across the body diagonal—that is, between points A and B. (**Hint**: Imagine a voltage V is applied between A and B, causing a total current I to flow. Use symmetry arguments to determine the current that would flow in branches AD, DC, and CB.)

A moderately difficult puzzle. This classic problem is valuable for several purposes. This is probably the best example of a network that cannot be solved by stepwise reduction of series and parallel combinations (the Wheatstone bridge looks easier, but the solution via Kirchhoff's laws is too time consuming to be convincing). Many students become enamored of those methods and must be convinced that they are not always ideal. This is also a good opportunity to demonstrate the power of symmetry arguments, which often are forgotten once the course leaves electrostatics behind. Furthermore, working the problem as described in the hint—that is, $(I/3)R+(I/6)R+(I/3)R = (5/6)IR$ so $V = I(5R/6)$ so $Req = (5/6)R$—helps give students deeper insight into the meaning of equivalent resistance: It is the ratio of V to I for a circuit rather than "the resistor I could have used instead." For talented students, extensions include the equivalent resistance across a face diagonal, along an edge, and across networks based on other platonic solids.

7. The figure shows a flexible wire loop lying on a frictionless table. The wire can slide around and change its shape freely, but it cannot leave the table surface. If a current flows in the clockwise direction in the loop, which of the following will occur?
   a) The loop will not move in any way.
   b) The loop will expand outward until it becomes a circle.
   c) The loop will contract down into a tangled mess.
   d) There is not enough information to tell.

A relatively easy puzzle. Many students will realize that as the current flows in each "subloop," it produces repulsive forces so that, on average, the largest forces will be the repulsive ones, causing the loop to expand. How well they state this argument will vary. Some students will argue from the point of view of given elements along the wire interacting with the "approximately dipole" field of the whole distribution. Again, they arrive at the correct result with varying levels of clarity. Nice extensions result from considering the same problem with applied fields in various directions.

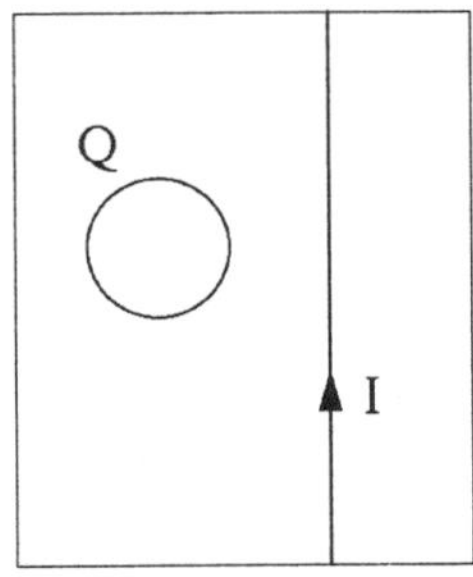

8. The figure shows an insulating ring and an infinite straight wire resting on the surface of a plane. The wire is fixed in position, but the ring may slide without friction on the surface of the plane. The ring is uniformly charged, with total charge $Q = 3$ nC. Initially, the wire carries a constant current of 5 A toward the top of the figure, and the ring is stationary. At $t = 0$, the current in the wire is abruptly reduced to 2 A. Which of the following best describes the motion of the ring after the current is reduced?

a) The ring will spin clockwise but remain stationary.
b) The ring will spin counterclockwise but remain stationary.
c) The ring will spin clockwise and slide toward the wire.
d) The ring will spin counterclockwise and slide toward the wire.
e) The ring will spin clockwise and slide toward the top of the figure.
f) The ring will spin counterclockwise and slide toward the top of the figure.
g) The ring will spin clockwise and slide away from the wire.
h) The ring will spin counterclockwise and slide away from the wire.
i) The ring will slide toward the wire without spinning.
j) The ring will slide toward the bottom of the figure without spinning.
k) none of the above (please describe what does happen)

A relatively difficult puzzle. The changing flux produces an ***E*** *field that is parallel to the wire, directed toward the top of the figure, and decreases in magnitude with distance from the wire. Thus, the ring experiences a net force toward the top of the figure and a net torque out of the plane of the page. As the ring gains angular velocity, it also experiences a net force toward the wire due to the Lorentz force, which is largest near the wire, where it is attractive. In addition to the complete answer, we give students credit for either answer d or f if they include the appropriate subsections of the argument here. The only extension we have ever ventured here is to use a negatively charged ring. Usually, the basic problem is enough.*

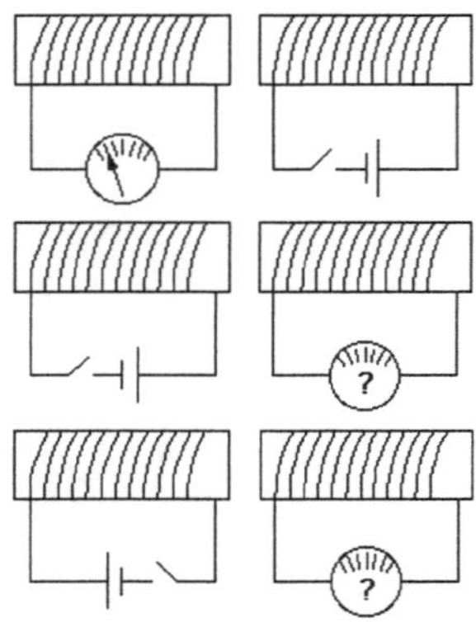

9. The figure above shows three "experiments" that may be performed with a pair of coils, a battery, an ammeter, and a switch. The ammeter is of the old analog type, so it reads positive current vs. negative current by deflecting to the left or to the right. In the first experiment, the coils are placed side by side as shown in the first pair of figures, and the switch is closed. At first, the ammeter deflects to the left; then it slowly drifts back toward zero. In the second experiment, the locations of the two coils are reversed before the switch is closed, as shown in the second pair of figures. In the third experiment, the coil with the battery is flipped over before the switch is closed.

For the second and third experiments, state whether the ammeter deflects to the left or the right when the switch is first closed.

A relatively easy puzzle. Many students figure this one out by looking carefully at the way the coils are wrapped to determine the direction of the field in the first coil rather than taking the first "experiment" as data to determine the result of the other two. It is useful to discuss the problem in class with no visual indication of the winding directions and show that the result is still obtainable.

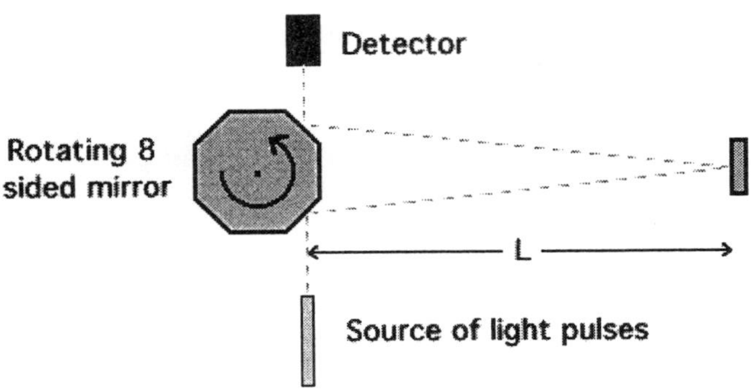

10. The sketch above shows the basic elements used by A. Michelson in the 1920s to measure the speed of light. A pulse of light hits the rotating eight-sided mirror and takes off on a round-trip to a flat mirror some distance off. The pulse will be received by the detector if the rotating mirror is in the right orientation when the pulse returns. A moderately priced power drill can turn at about 1200 RPM. If such a drill is used to turn the mirror, how far away does the flat mirror have to be for a pulse to be observed? How about for a Dremel that turns at 30,000 RPM?

(**Hint**: Assume the distance to the mirror (L) is large, so you can approximate the round trip as exactly $2L$.)

A relatively easy puzzle. Most students will either get this one correct or be off by a factor of 8. Some will calculate correctly but lack confidence in their answer due to the very long baseline required (Michelson did this experiment between two mountaintops in the Rockies). Extensions include doing the experiment underwater and other mechanical methods to determine c (e.g., using a chopper-wheel assembly).

Optics

1. Calculate the minimum size of a mirror such that a six-foot-tall person can see himself from head to toe. Also, state how far above the floor the bottom of the mirror must be mounted.

A relatively easy puzzle. Many students will approach this by guessing a specific distance to the mirror. Some will notice that this parameter drops out, and others will not. Discussing this point is always valuable, and some students will vehemently deny that it is true! I usually encourage these students to use a dry-erase type marker and test the theory on a full-length mirror in one of the university washrooms. Students are also amazed that the length needed is only half the height. Students will also display varying levels of sophistication on the issues of where to place the mirror. We give credit if they specifically state that they are assuming eyes at the top of their heads, but we encourage estimates of a few inches by highlighting these answers in class.

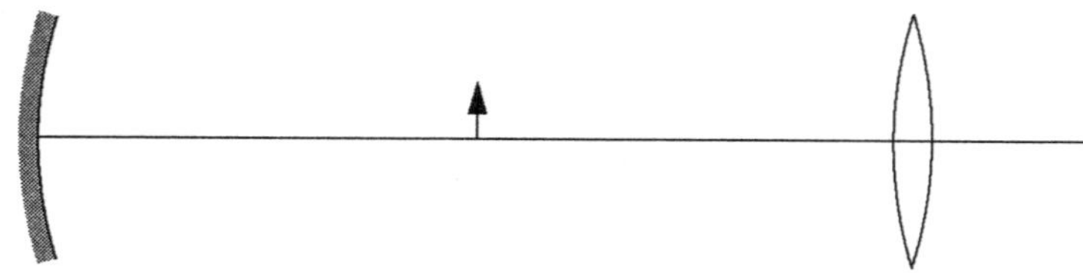

2. The figure above shows a concave mirror with $f_M = 12.5$ cm and a converging lens with $f_L = 25$ cm. They are placed 50 cm apart with an object centered between them. Describe the image or images created by this apparatus.

A moderately difficult puzzle. Students that rely on ray-tracing methods will believe the image is due to the mirror alone: inverted and at the location of the object. However, the image at infinity produced by the lens is very difficult to obtain for inexperienced ray-tracers. This can lead to a nice discussion of the meaning of images (and objects) at infinity, and of "special points" for mirrors and lenses. Furthermore, students find it interesting to discuss the application of this "device" in lighthouses.

3. Suppose you place a transparent decal in the rear window of your car so that the driver behind you can read it normally. What will you see if you look at it in your rearview mirror?

A relatively easy puzzle. Many students will determine that the decal is "normally" readable by trying it out. This puzzle is the best way we have found to broach the very confusing topic of front-back reversal by mirrors. Extensions include discussions of kaleidoscopes, halls of mirrors, image inverters for telescopes, and other multiple-reflection devices.

4. A person with good vision finds that she cannot focus on anything underwater. However, plastic goggles with "lenses" that are **flat** plastic disks allow her to see the fish clearly. Please explain how this can be. (**Note**: It does not have to do with salt water or chlorine irritating her eyes.)

A relatively difficult puzzle. This puzzle provides a good introduction to the whole question of lens making, and ties the lens maker's equation back to Snell's law. It should not be difficult, but many students have a hard time seeing where the "real question" is. Discussing this question helps bring home the point that the human eye is best viewed as the last refractive element in an optical system rather than as a generic "receiver" that interprets the results of external optics.

Chapter 10: Physlet Problems

General Principles

The rapid pace of hardware and operating-system development has made it difficult for educators to produce computer-rich curricular material that was not obsolete shortly after publication. The half-life of a typical computer desktop was shorter than the textbook publication cycle, and this all too often led to minimal documentation, poorly tested nonstandard user interfaces, and idiosyncratic behavior. Computational physicists accustomed to programming in Fortran had little interest in education and user-interface design. The tools used by scientists in day-to-day office work such as correspondence, class management, and professional publication interoperated poorly with programming tools and with educational software. Publishers were therefore reluctant to integrate computer usage into primary educational texts, and the mechanism to effectively create and distribute media-rich documents from the desktop simply did not exist.

It is not surprising that many thoughtful teachers were unwilling to invest the time and energy to overcome these obstacles. Since little research had been done on the effectiveness of most educational software, this wait-and-see approach may have been wise. However, a case can now be made that this throwaway cycle for educational software need not repeat itself; key technologies are now available that enable authoring, evaluation, and distribution of curricular material that will withstand the test of time. The most promising technologies are based on virtual machines, meta-languages, and open Internet standards. These technologies are platform independent. Manufacturing, inventory control, and advertising have, in effect, provided the education community with a rich and flexible set of standards to enable electronic curriculum distribution.

The Java programming language, coupled with a scripting language such as JavaScript, is widely used on the Internet and is likely to play a leading role in developing computer-rich curricular material. We have adopted this technology to develop and deliver interactive Physlet problems to our students.

Physlet problems are ideally suited for media-focused WarmUps and Puzzles, since this Web-based technology can provide interactivity that is difficult to produce with static text. Not only do these problems present new challenges to students but "correct" answers can often be obtained using numerous problem-solving techniques, thereby providing an opportunity for class debate. This chapter provides examples of effective Physlet-based problems. Each problem begins with a screen shot of the embedded applet, a "screenplay" describing what the user sees, and possible user interactions. We include text for one or more problems based upon the screenplay, and an answer.

Kinematics

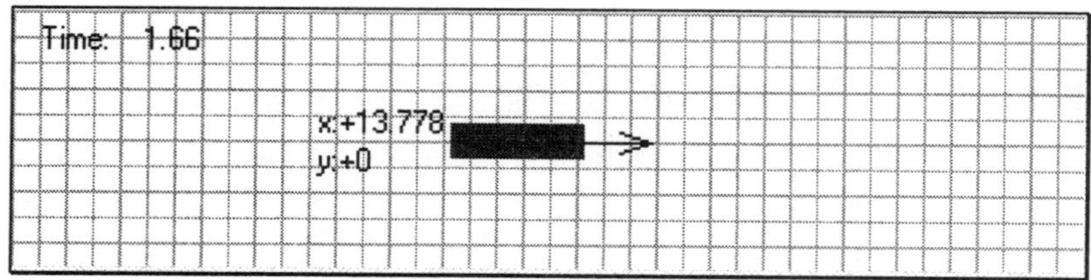

Physlet Problem 1. *The rectangle starts at rest at t = 0 and moves to the right with increasing velocity. The acceleration of the rectangle is constant. The black arrow moves with the rectangle and has a constant length. The simulation runs from t = 0 to t = 2 and repeats. The student cannot control this applet in any way.*

Question 1. Physicists use arrows to represent many things in diagrams. What vector quantity is being represented by the arrow in this simulation?
a) displacement
b) velocity
c) acceleration
d) speed

Answer: acceleration. This problem can be given before acceleration is covered in class, i.e., as a WarmUp. Students can eliminate the three incorrect answers since they are clearly not constant.

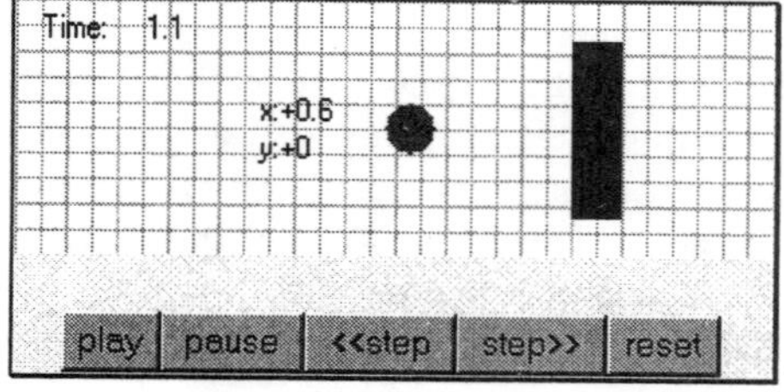

Physlet Problem 2. The animation begins with the puck at $x = -6$ at $t = 0$. The puck then moves to the right at a constant speed of 6 m/s until it bounces off of the wall at $t = 2$, after which it moves to the left with a constant speed of 4 m/s until the simulation stops at $t = 5$ s.

Question 2a. A hockey puck bounces off a wall and returns to its original starting point, where it is stopped as shown in the animation. What is the average speed of the puck during the time it is in motion (position is shown in feet, and time is in seconds)?

Answer: 4.8 ft/s

Question 2b. A hockey puck bounces off a wall and returns to its original starting point, where it is stopped as shown in the animation. What is the average velocity of the puck during the time it is in motion?

Answer: 0.0 ft/s

Question 2c. A hockey puck bounces off a wall and returns to its original starting point, where it is stopped as shown in the animation. What is the instantaneous velocity of the puck at $t = 3$ s?

Answer: –4.0 ft/s

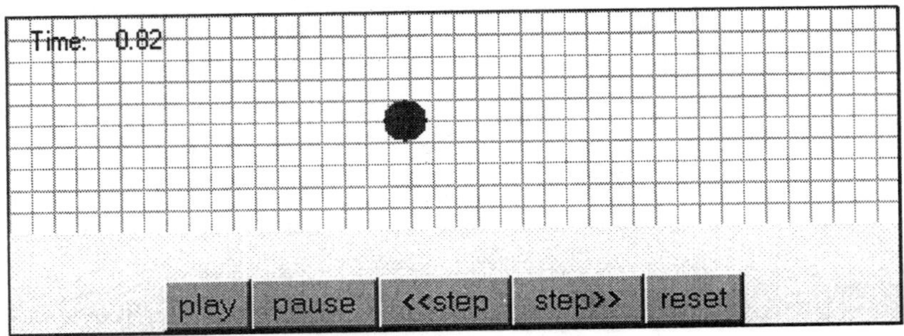

Physlet Problem 3. This is one of our most effective kinematics problems. A ball moves across the screen with constant acceleration. Values near +/– 4.5 units/time^2 are particularly effective. The animation should start at time $t = 0$ with the trajectory set so that the ball is off the left- or right-hand side of the screen. That is, the ball should not be visible at $t = 0$ and should move onto the grid at some later time. Many students simply cannot do a problem that does not give them an easy way to determine a value for the initial velocity, v_0. This problem should be given with both positive and negative acceleration since students tend to associate the direction of motion with the direction of the acceleration.

Question 3a. What is the acceleration of the ball? You may click-drag the mouse inside the animation at any time to measure position.

*Answer: Determined by the script. For example, the script document.Animator.addCircle(5,"-40 + 4*t - 2 t*t", "0"); will display a ball with an acceleration of 4 units / time^2. We have observed that some students will mindlessly give an answer of 9.81 m/s^2 if the problem is recast to show vertical motion.*

This problem is an effective tool for discussing measurement error since a straightforward application of $v = \Delta x/\Delta t$ will likely give incorrect values due to the difficulty of measuring the center of the ball using the mouse. Even good students can disagree about the correct result. Students should be required to obtain an answer that is correct to better than 2 percent.

Question 3b. What is the average velocity from the time $t = 1$ to $t = 3$? What is the average acceleration? You may click-drag the mouse inside the animation at any time to measure position.

Answer: Determined by the script as in 3a. Different students may be assigned different time intervals in order to make the point that instantaneous and average acceleration have the same value if the acceleration is constant.

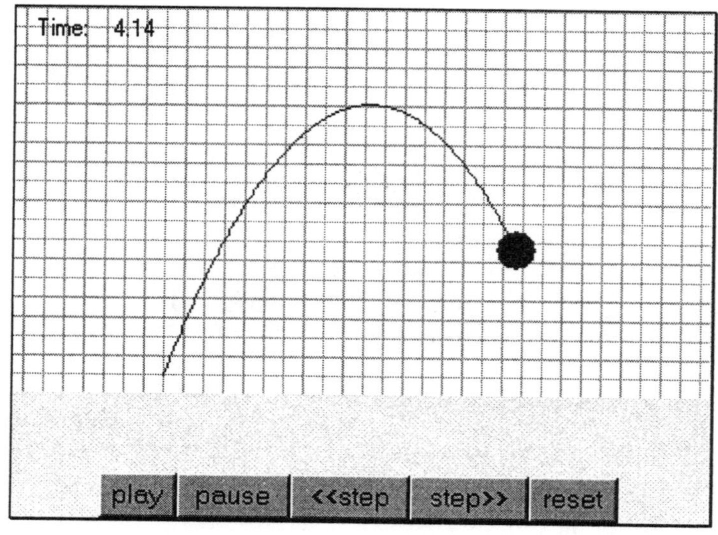

Physlet Problem 4. A ball moves across the screen in projectile motion, leaving a visible trail to highlight its trajectory. Students may click-drag the mouse to measure coordinates. It is difficult to determine the velocity by measuring $\Delta \mathbf{r}/\Delta t$ using a single time step because of the one-pixel limit imposed by the screen resolution.

Question 4. What is the minimum speed of the projectile? You may click-drag the mouse inside the animation at any time to measure position.

Answer: Since the acceleration of the ball is constant in the –y direction, the projectile has its minimum speed at the maximum height when the y component of its velocity is zero. The x component of the velocity is constant and is easily measured by taking the distance traveled during a 5-second interval and dividing by time. This approach may not be obvious to students.

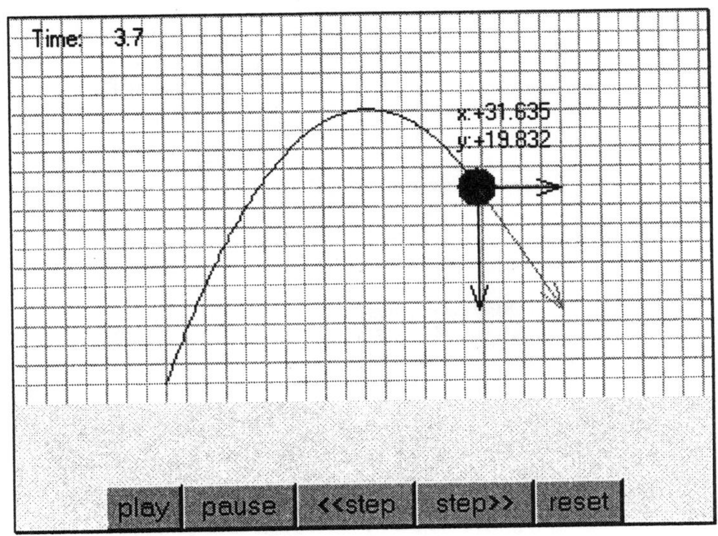

Physlet Problem 5. This animation presents a more elaborate version of Physlet Problem 4. The velocity components and particle position are shown as the animation progresses. This example demonstrates how it is possible to modify a script in order to add layers of complexity as students show mastery of basic concepts.

Question 5. What is the speed of the projectile at $t = 3.7$ s? What is the velocity of the particle? The acceleration is constant but not necessarily 9.8 m/s^2. The scale used to display velocity vectors is arbitrary. That is, do not assume that a vector of length five represents a velocity of 5 m/s.

Answer: Since x and y are displayed to five significant figures, it is possible to obtain a good value of the velocity at t = 3.7 s using the definition $\Delta \mathbf{r}/\Delta t$. *A number of students will attempt this problem by finding the initial velocity and the acceleration in order to use the constant acceleration kinematics equations given in the text.*

Mechanics

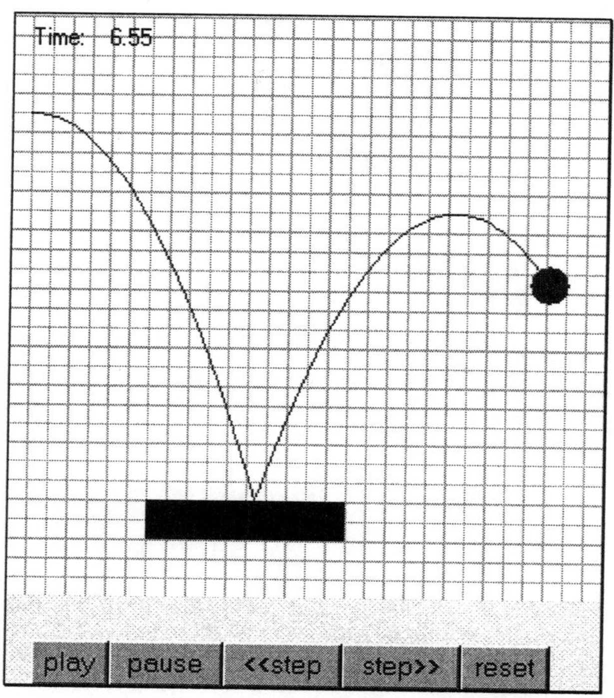

Physlet Problem 6. A ball starts with an initial horizontal velocity, falls, and bounces off an immovable black rectangle, leaving a trail to mark its trajectory.

Question 6a. A 100-g ball is thrown and bounces off of a solid object as shown in the simulation. What impulse is delivered to the ball by the collision? The vertical acceleration is -9.81 m/s^2. Neglect air resistance.

Answer: Determine the impulse by first determining the change in the vertical component of the velocity. This is most easily done using the conservation of energy and the two maximum heights in the trajectory.

Question 6b. A ball is thrown and bounces off of a solid object as shown in the simulation. Determine the coefficient of restitution of the ball. The vertical acceleration is -9.81 m/s^2. Neglect air resistance.

Answer: Since the coefficient of restitution is the ratio of the energy before the collision to the energy after the collision, it is equal to the ratio of the heights of the two trajectory maxima.

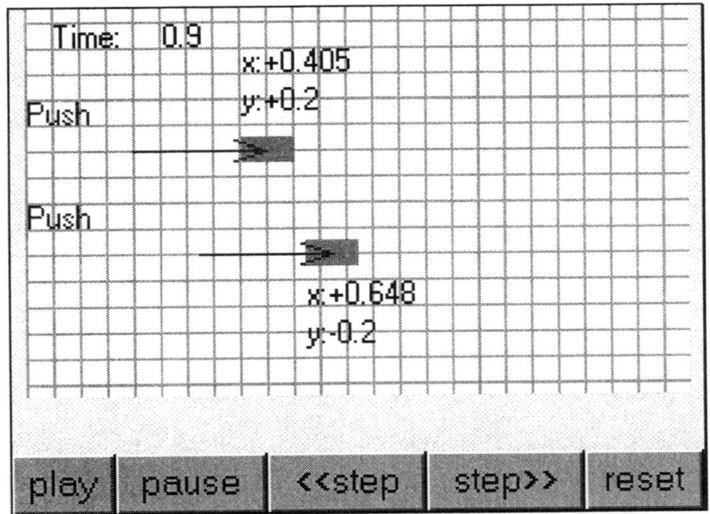

Physlet Problem 7. The animation shows two carts accelerating from rest with $x = 0$ at $t = 0$. The acceleration stops at $t = 1$s after which time the carts move at constant velocity. Force vectors are shown during the time they are accelerating. This mastery problem tests a number of concepts in Newtonian mechanics. Students are surprised that the heavier cart receives less kinetic energy for the same amount of push.

Question 7. Identical forces push two carts for 1 second. Neglect friction.
Determine the ratio of the mass of the green cart to that of the red cart.
Determine the ratio of the momentum of the green cart to that of the red cart.
Determine the ratio of the kinetic energy of the green cart to that of the red cart.

Answers: (a) 1.6:1 (b) 1:1 (c) 1:1.6

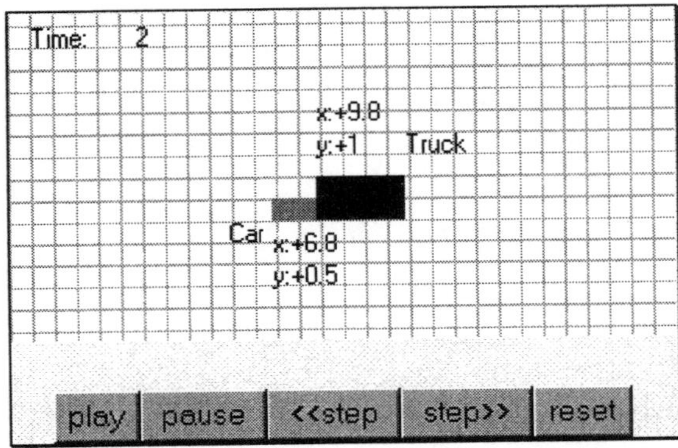

Physlet Problem 8. Two rectangles start from rest and accelerate to the right.

Question 8. A 4500-kg truck breaks down and receives a constant push by a 1500-kg car as shown in the animation (position is in meters, and time is in seconds). What is the force of the car on the truck?

Answer: The net force on the truck is its mass x acceleration. Acceleration is determined from the kinematics equation $x(t) = (1/2)a\, t^2 + x_0$ by measuring x_0 at $t = 0$ and one other position at a later time. Since the only force acting on the truck is the push of the car, the force of the car is the same as the net force on the truck.

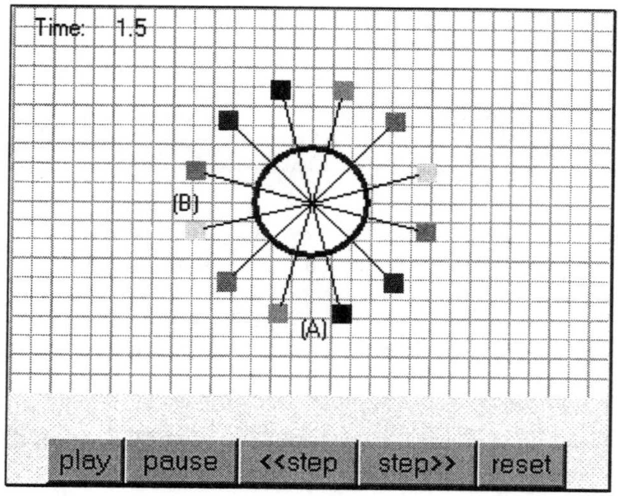

Physlet Problem 9. A Ferris wheel is simulated by a collection of colored squares moving along a circular path at constant angular velocity. Users may click-drag the mouse at any time to measure position.

Question 9. A Ferris wheel rotates at constant speed as shown in the animation. Each square represents a chair on the Ferris wheel. What applied forces act on a rider when the rider is at point (A), $(x, y) = (0, -6)$? At the point (B), $(x, y) = (-6, 0)$?

Answer: Since the rider is in uniform circular motion, the net force on the rider is always the centripetal force, mv^2/r.

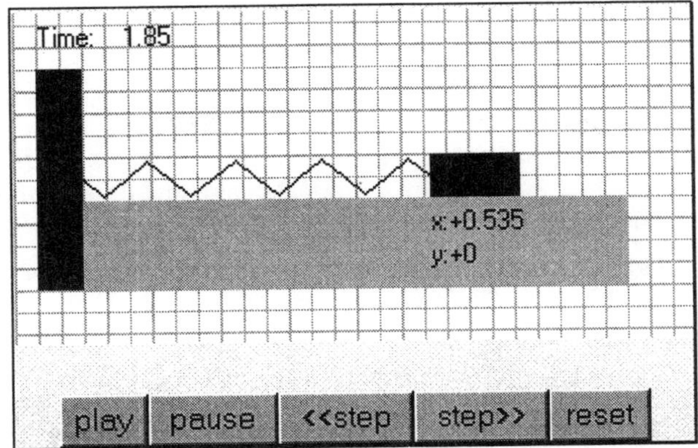

Physlet Problem 10. A rectangle oscillates horizontally with a period of 2 seconds.

Question 10. A 0.5-kg cart resting on an air track oscillates as shown in the animation, with position in meters and time in seconds. What is the spring constant of the spring?

Answer: The most accurate solution to this problem is obtained by measuring the time for 10 oscillations and then using $\omega^2 = k/m$. *Conservation of energy can also be used effectively since the velocity at the equilibrium point can be determined accurately from the high precision coordinates of the mass.*

Waves

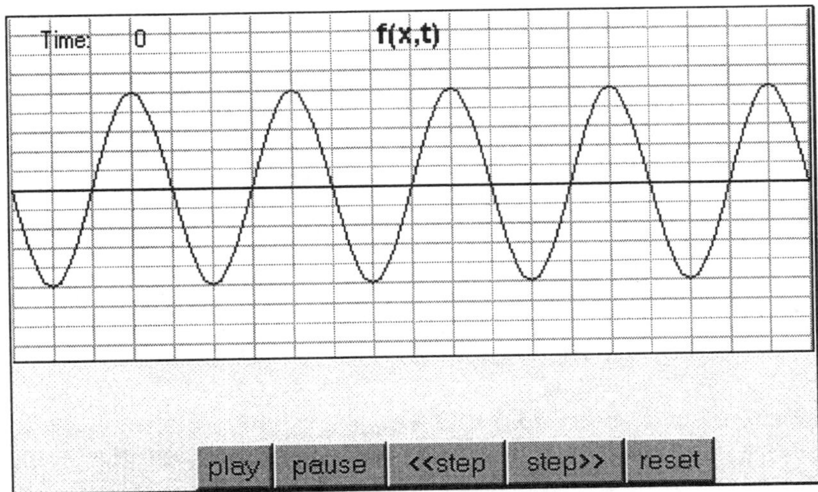

Physlet Problem 11. The animation shows a standing wave with a period of 2 time units and a wavelength of 4 distance units. Click-dragging the mouse allows the user to measure coordinate values.

Question 11. The above simulation shows a standing wave on a string. With what speed do waves propagate on this string? Assume that time is measured in seconds and distance is measured in meters.

Answer: Although the animation shows the string moving transverse to the x axis, the speed of the wave is not observed directly but must be derived using $v = \lambda f$.

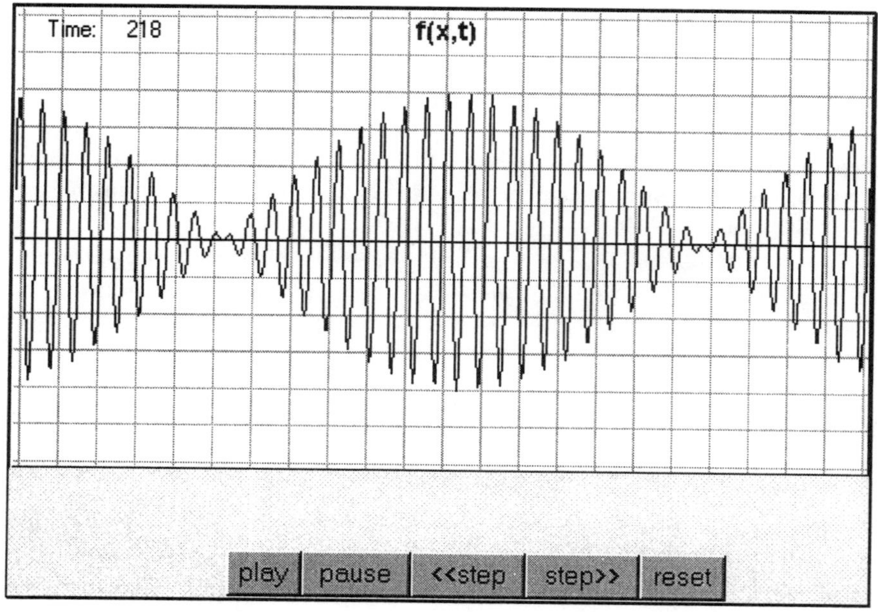

Physlet Problem 12. The animation shows a traveling wave propagating to the right.

Question 12. The above simulation shows a traveling wave. Assume time is measured in milliseconds and distance is measured in meters. Do not assume that this wave is propagating in air. You may pause the simulation at any time and click-drag the mouse to measure coordinate values. Describe the sound that is heard.

Answer: The listener will hear a beat frequency of 5 Hz superimposed on an average tone of 100 Hz. Students often confuse amplitude vs. time graphs with amplitude vs. space graphs.

Electricity and Magnetism

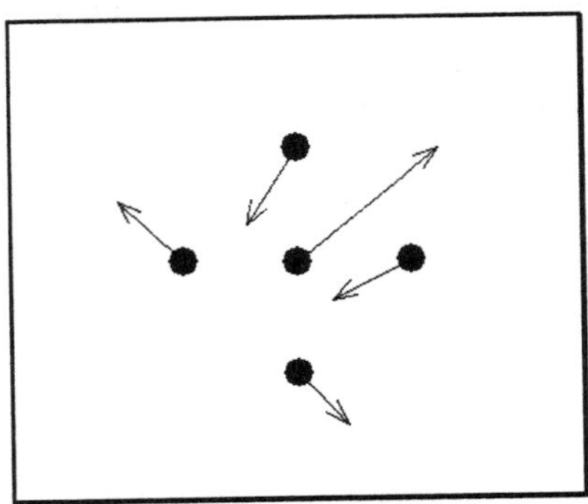

Physlet Problem 13. Five objects are shown on the screen along with vectors that are proportional to the net force on each object. Vectors are redrawn if any object is click-dragged to a new position. Notice that the inverse square law allows one to determine the nature of the force between any two charges, i.e., attractive or repulsive, if they are brought close together.

Question 13. How many charges have like signs? You can click-drag on any charge to change its position.

Answer: Three charges have one sign, two charges have the opposite sign. This is observed by selecting an arbitrary charge as a test charge and dragging it close to each of the remaining charges in order to observe the direction of the force.

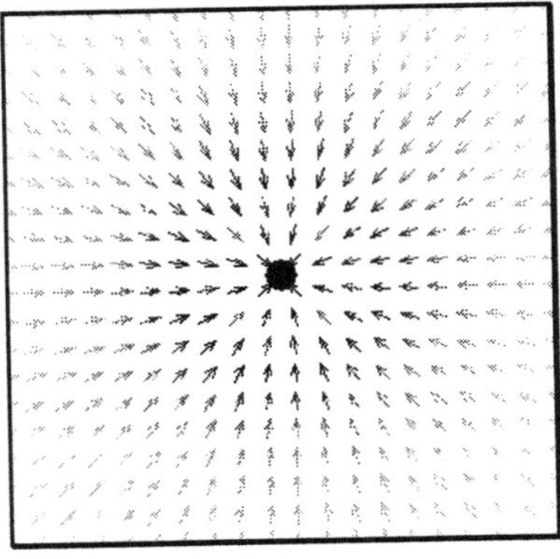

Physlet Problem 14: A vector field is shown using color to represent magnitude. When a user click-drags the mouse inside the Physlet, a yellow message box appears showing the magnitude of the field at the mouse position.

Question 14a. An unknown vector field is shown. The vectors shown point in the direction of the electric field, and the color represents the field's magnitude. Click-drag to measure the field at any point. Is the field a Coulomb field from a point charge? If so, find the charge. If not, can you think of another charge distribution that would produce the field shown?

Answer: Depends on the script. This force field shown above is not a point charge, because it does not decrease as the inverse square of the distance $1/r^2$. Click-dragging will reveal that this field has a $1/r$ dependence, so the simulation could represent a line charge perpendicular to the plane of the applet.

Question 14b. The above simulation represents either a point charge or a line charge perpendicular to the screen. Determine which type of charge distribution is being represented. The vectors shown point in the direction of the electric field, and their colors represent the field's magnitude. Find either the magnitude of the point charge or the density of the line charge by click-dragging and reading values from the yellow message box. Assume distance is measured in millimeters and electric field is measured in N/C.

Answer: Depends on the script. Does the field decrease in proportion to $1/r^2$ or $1/r$? Measure the field and use $|E| = kq/r^2$ if the field is a point charge or $|E| = 2k\lambda/r$ if the field is a line charge.

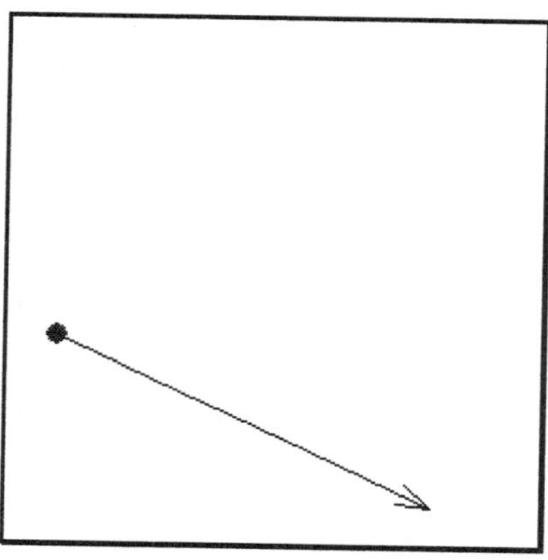

Physlet Problem 15. A red test charge can be moved inside the applet with the mouse. The blue vector attached to the test charge represents the direction of the force on the charge. A yellow message box appears in the lower left-hand corner of the Physlet, showing the coordinates and the magnitude of the force on the charge.

Question 15. One or more objects with uniform charge are located just off the screen on the left-hand side. Use the test charge to measure the electric field, and from that determine what types of charges are present.

Answer: Depends on the script. Students should be able to identify a single point charge, two point charges with like and unlike signs, and a line charge by examining the magnitude and direction of the force on the test charge.

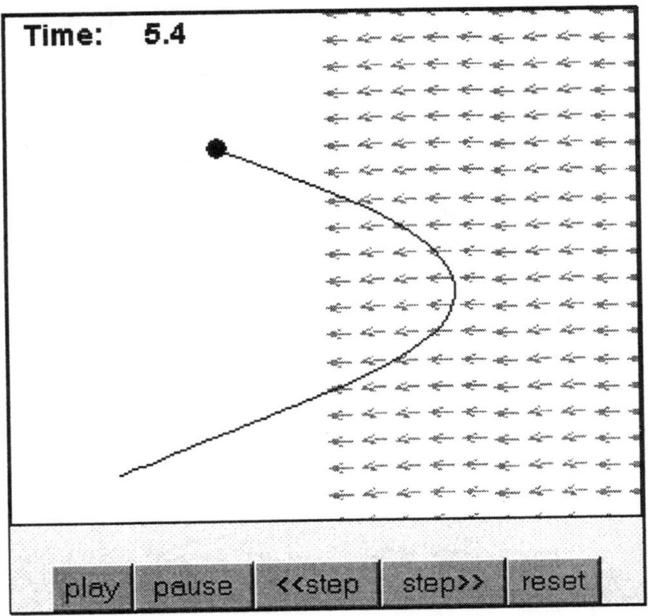

Physlet Problem 16. A black ball moves at a constant speed until it encounters a constant vector field that changes its direction. Time is shown in the left-hand corner of the simulation. Position can be measured by click-dragging the mouse.

Question 16. A 1-mg particle with a charge of 2 μC is fired into an unknown electric field as shown above. Find the magnitude of the electric field. Click-drag to measure the position in meters. Time is measured in seconds.

Answer: Depends on the script. This simulation has an electric field of magnitude 1.5 N/C.

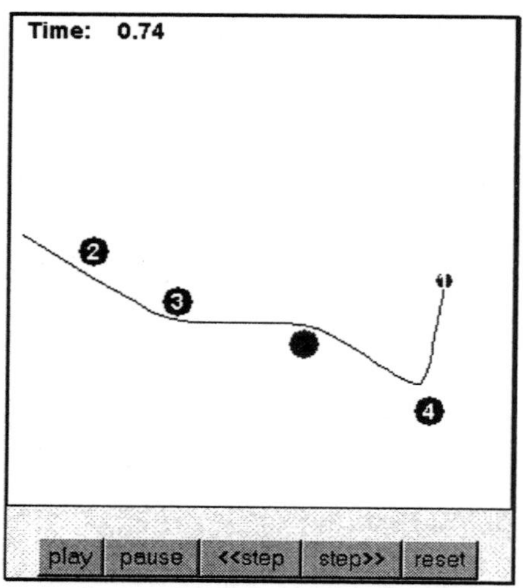

Physlet Problem 17. A blue test charge moves under the action of the electric field created by four fixed charges. Its trajectory is drawn as a blue line as it moves.

Question 17. A test charge is fired past four fixed charges as shown in the simulation. The red charge is positive. Determine the signs of the unknown charges. You may consider neutral to be a possible answer.

Answer: The test charge is negative since it is attracted toward the positive charge. Charge 2 is neutral since it does not deflect the test charge. Charge 3 is positive since it attracts the test charge. Charge 4 is negative since it repels the test charge.

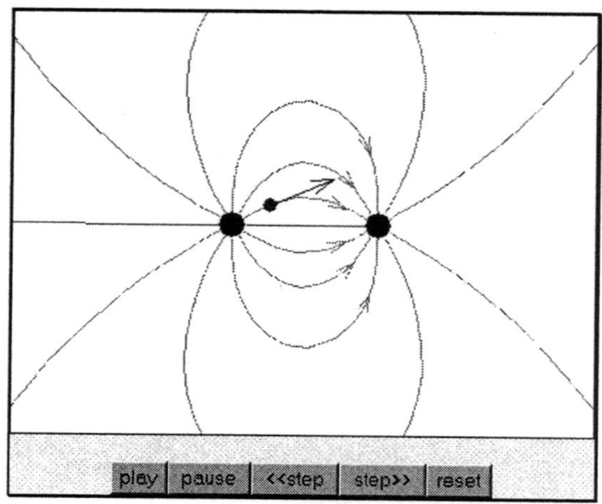

Physlet Problem 18. Two fixed charges are shown as black circles along a draggable test charge drawn in red. The dark blue arrow attached to the test charge represents the force on the test charge. The light blue lines represent the field lines. When the Play button is pressed, the test charge accelerates due to the electric field. The test charge marks its trajectory with a red line as it moves under the influence of the fixed charges.

Question 18a. Click-drag the test charge and describe the relationship, if any, between the dark blue force vector and the light blue electric field lines.

Answer: The force vector is always tangent to the electric field lines. The force is larger wherever there are many field lines and smaller wherever there are few field lines.

Question 18b. Place the test charge on a field line and start the animation. Does the trajectory follow the electric field lines? Explain.

Answer: No, the trajectory does not follow the field lines. The field lines show the direction of the force on the positive test charge.

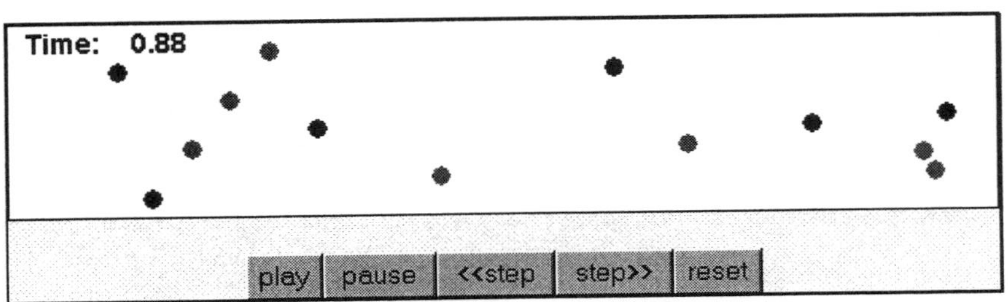

Physlet Problem 19. Blue and green spheres move to the right and left, respectively. Each sphere represents a charge, and the color signifies the sign of the charge.

Question 19. A drift tube is shown above. The green dots represent atomic clusters of charge 1 nC and the blue ones clusters of charge –2 nC. What is the total current to the right?

Answer: Count the number of positive charges entering the drift tube from the left, N_1, and the number of negative charges entering from the right, N_2, during a 1 s interval. The total current can be calculated using:

$$I = Q_1 N_1 + Q_2 N_2 .$$

Students will often make a sign mistake since negatively charged particles traveling to the left produce a positive current to the right.

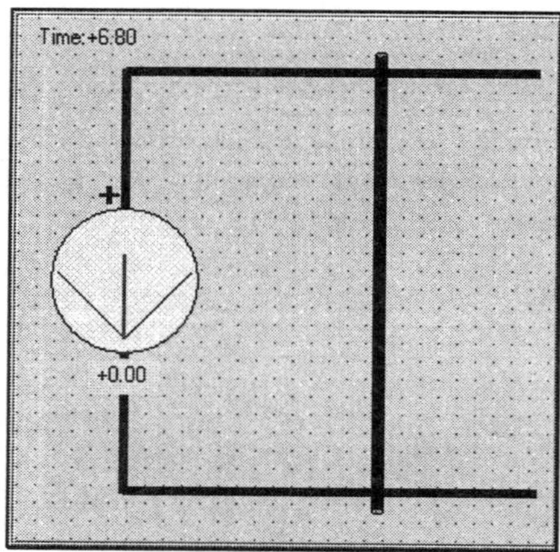

Physlet Problem 20. A circuit is shown consisting of a fixed green U-shaped wire and a movable wire drawn in black. The U-shaped wire includes a galvanometer. A magnetic field into or out of the plane can be specified as a function of x.

Question 20. Click-drag the black wire and observe the galvanometer. Remember that current flowing into the + terminal, i.e., counterclockwise, will deflect the meter to the right. Describe the magnetic field passing through this circuit.

Answer: Determined by the script. For example, the script

*document.Faraday.setFieldFunction(" 0.2*x ");*

simulates a magnetic field out of the plane that increases linearly as x increases.

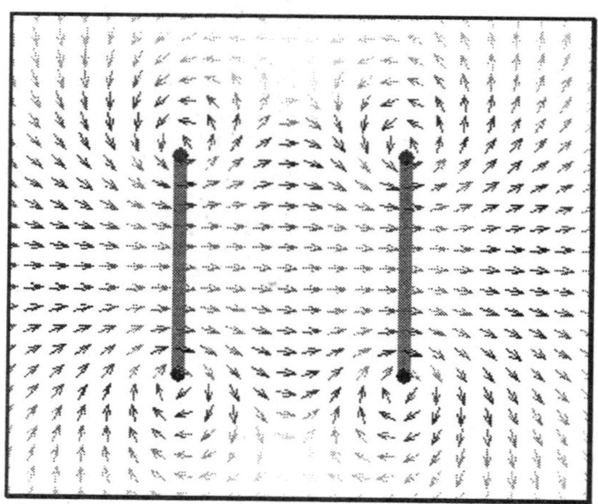

Physlet Problem 21. The Faraday Physlet displays a cross-section of two current carrying coils and their associated magnetic field. The user can move a coil by click-dragging near the coil center. The user can also change the radius of a coil by click-dragging the top or bottom of the coil. Numerical values are displayed in a yellow message box during mouse operations.

Question 21. Click-drag the top of each coil so that its radius is set to 4. Adjust the separation of the coils to produce a Helmholtz configuration. Determine the current in the coils by measuring the magnetic field near the center of the coils. Over what range of y does the magnetic field vary by less than 10% of the value at the center? Assume position is measured in meters and magnetic field is measured in mT.

Answer: The field is uniform to within 10% for $\Delta y = 1.6$ m about the center.

Optics

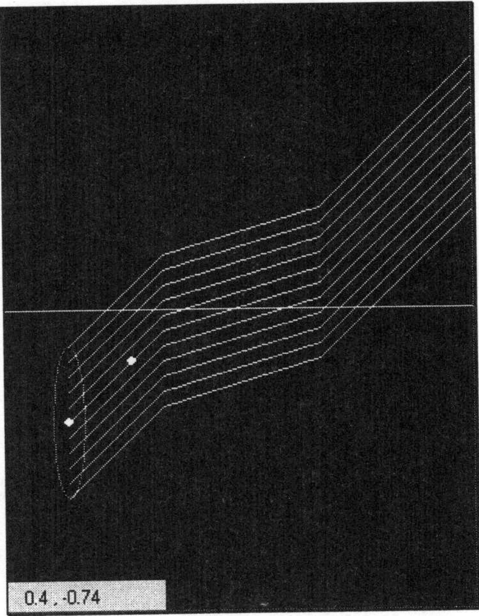

Physlet Problem 22. A beam of parallel rays propagating to the right is refracted upon entering and leaving another medium. Both the position of the source and the angle of the source can be adjusted by click-dragging the circular hotspots. Angles may also be measured by click-dragging away from the hotspots.

Question 22a. A light source emits a beam as shown in the simulation. What is the index of refraction of the substance? You can click-drag both the position and the ray-angle for this source.

Answer: n = 2.4

Question 22b. Click-drag the light source into the region of higher index of refraction. Adjust the angle of the source in order to demonstrate total internal reflection. What is the angle of total internal reflection?

Answer: θ = 24 degrees

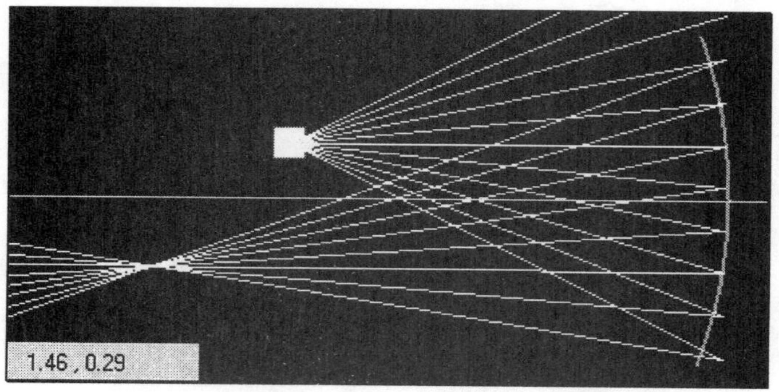

Physlet Problem 23. A point source emits a cone of light that eventually strikes a mirror. The position of the source may be adjusted using the mouse.

Question 23. A point source is located to the left of a mirror. You can drag this point source to any position. Find the focal length of the mirror.

Answer: f = 1.3 m. This problem is easy if the source is dragged onto the optic axis and its position adjusted so that the incident and the reflected rays overlap. This occurs when the source is at the radius of curvature of the mirror.

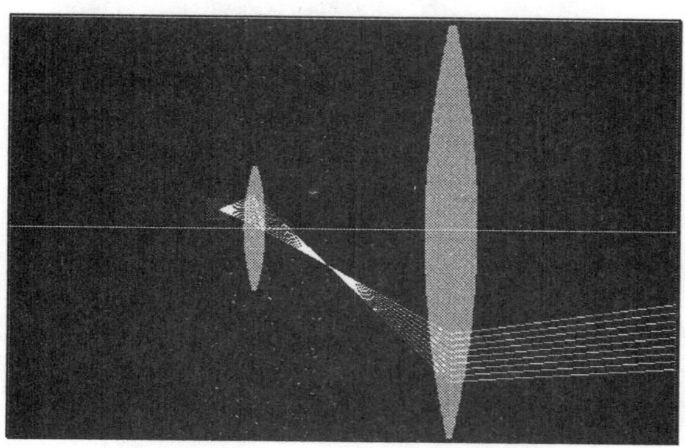

Physlet Problem 24. A point source emits a cone of light before it passes through two lenses. The position of the source may be adjusted using the mouse.

Question 24. Two lenses, an eyepiece and an objective, are used to make a microscope. Where should the object be placed for optimal viewing by a relaxed eye? You may focus the microscope by click-dragging the object into position. Position is measured in cm.

Answer: The microscope is in focus when the light rays leaving the second lens are parallel. This corresponds to the object being at infinity, the optimal distance for viewing by a relaxed eye. Many students find this result confusing since they assume objects at infinity must appear very small.

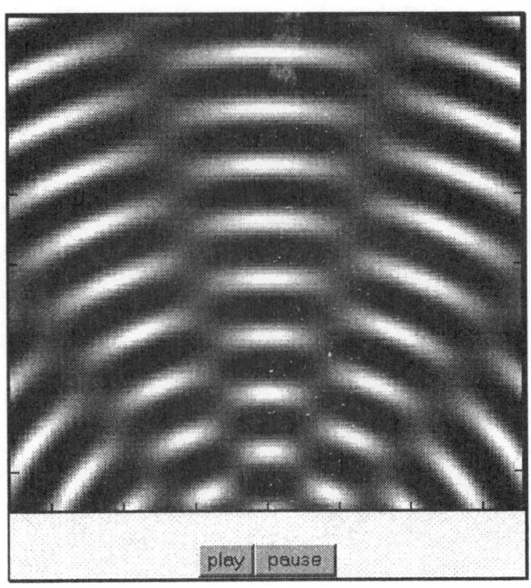

Physlet Problem 25. A ripple-tank wave pattern is created and displayed whenever the "play" button is pressed. The mouse can be used to make coordinate measurements.

Question 25. A double slit is hidden somewhere below the animation. What is the slit separation? You can measure (x, y) coordinates by click-dragging inside the animation. Assume all measurements are given in nanometers (nm).

Answer: 2 nm

Chapter 11: Scripting Physlets

Basics

Modern markup languages such as Hypertext Markup Language, HTML, make it possible to create multimedia-enhanced documents in a platform-independent fashion. Since these documents are text documents, they can be prepared with most text editors, given to students on a floppy or transmitted via the Internet, and viewed with standard desktop applications such as word processors and Internet browsers. Yet, the HTML aware application displays full multimedia information, including text, graphics, video, and sound. The recent introduction of the Java programming language by Sun Microsystems now makes it possible to add platform-independent programs to this multimedia stew. Java accomplishes this trick by specifying a relatively simple Virtual Machine (VM), which can be implemented on any computer architecture, i.e., Unix, Mac, or Windows. Although this VM does not provide as rich a set of tools as the native operating system, the virtual machine can have a user interface with buttons, a drawing canvas, and other graphical elements. There is virtue in simplicity. Small platform-independent programs are ideally suited for instructional purposes such as homework problems, class demonstrations, or JiTT. Applets that have been embedded into HTML pages can interact with the user via a scripting language such as JavaScript. For simplicity, we have named small scriptable applets capable of displaying physics content "Physlets."

A good collection of over 250 Physlet-based problems can be found on the Prentice Hall Companion Web site to Douglas Giancoli's algebra-based textbook *Physics: Principles with Applications*. The interactive problems used on this site are based upon a dozen of these Physlets. Each problem relies on the appropriate script to customize the Physlet's behavior for that particular problem. Although the problems on the Prentice Hall site are protected by copyright, Physlets are free for noncommercial use at educational institutions. Physics educators are encouraged to download Physlets from the Davidson College WebPhysics site and to write their own scripts. Physics education research notwithstanding, it is unlikely that a single teaching style will be adopted by—or be effective with—all physics teachers. Many instructors want to tinker with and modify even well-written problems in order to provide variety in their teaching, to adapt the problem to their own interests, and to meet the needs of diverse student populations. Subsequent sections of this chapter will demonstrate that embedded Java applets are one of the most promising technologies for achieving this flexibility.

Internet Technology

Web technology is still evolving, and it is currently not possible to run every Physlet on every computer platform using any browser. The promise of "write once, run everywhere" is unfortunately still a promise. However, there are powerful commercial reasons for adhering to a core set of Internet standards, and it is very likely that these standards will prevail. The Sun-Microsoft court battle over the Java trademark notwithstanding, Microsoft is supporting Java and has adopted HTML as the new Windows Help file format. Microsoft is producing a scriptable Java product to allow any Java-enabled application to connect to any standard ODBC database. (This idea is very similar to Physlets. Write a general-purpose Java applet for Internet commerce and script the applet to provide functionality needed for a particular application.) The European Union has adopted a specification for JavaScript, officially called ECMAScript, that browser vendors will be required to adhere to if they are to sell in Europe. There are numerous software packages that simplify high-level authoring using JavaScript that also check for browser compatibility. Hewlett-Packard, Novell, and IBM are making substantial investments in Java and are putting pressure on Sun to keep the standards process open. And finally, an enterprising group of developers with the Free Software Foundation is reverse engineering Java to provide a freely available version of the product complete with source code. Anyone who has ever written and distributed curricular material using proprietary authoring packages will appreciate the benefits to the education community of supporting and adhering to standards. Java and JavaScript are likely to be an excellent option for producing Web-based curricular material for the foreseeable future.

It is not necessary to become a Web expert in order to use Java applets in HTML documents. Many Java applets are available for free on the Web, and the process of downloading and embedding these applets is only slightly more difficult than embedding an image. However, just as there are various image file formats, there are various versions of Java and various ways of packaging the applet. Computers are notoriously unforgiving of syntactic errors, so it is important to have a bird's-eye view of Java technology in order to understand what may go wrong. You may want to skip ahead to the section on embedding and refer back to the remainder of this section when we discuss how Java Archive (jar) files are used to distribute Physlets.

Java has evolved rapidly since its introduction by Sun Microsystems in late 1995, and it has been difficult for software vendors to update their products to keep pace with new language features. Netscape, Microsoft, and other vendors quickly adopted the first version of Java, version 1.0, and applets written for this version will almost certainly work with any application that supports Java. This includes Netscape Navigator, Internet Explorer, and authoring packages such as FrontPage and PageMill. Unfortunately, Java 1.0 had serious limitations as a programming language and needed fundamental changes and extensions in order to reach its full potential. Although Java development environments from Borland, Symantec, Sun, and Microsoft quickly adopted the new specification, Internet browsers are only now beginning to support the new version. Running a Java 1.1 applet in a Java 1.0 aware application is likely to produce a gray box and a rather cryptic "Class not found" error message. Furthermore, Sun Microsystems and Microsoft have extended the language in different ways to provide access to native graphical user interface (GUI) components such as dialog boxes, toolbars, and menus. It is unlikely that the promise of platform independence is attainable in the near future if developers use some of the extensions available from competing vendors. However, the core language, including many version 1.1 features, is very stable, and most educational software developers have chosen to stick with the original, albeit more limited, GUI components available since Java 1.0. This seems like a small price to pay, and all Physlets have been written in "pure Java" to provide vendor and platform independence.

It is also helpful to understand the process of writing Java code in order to understand the file structure that authors will encounter when using Java applets. Preparing Java code that will eventually run inside a Java aware application is a multistep process. First write the code! Java syntax is similar to C in some respects, but its philosophy is much closer to Object Pascal or Smalltalk. The programmer begins by creating a file that has the applet name as the file name and has the extension "java." This main file often contains the code necessary to generate the user interface. It also contains code for various public methods that can be accessed from outside the applet using a scripting language such as JavaScript. (Methods appear to behave like subroutines or functions in procedure-oriented languages such as Basic or Fortran, but they are really quite different. See, for example, *Thinking in Java* by Bruce Eckel, Prentice Hall 1998.) Almost any significant software project will require additional source-code files. Applets that need additional Java files can access them by specifying their name next to the "import" keyword near the beginning of the file. This process is very similar to accessing a library routine in other languages. A number of files can be grouped together by adding the "package" keyword followed by a name as the first line of a Java program. These files are usually stored in a subdirectory having the same name as the package. Blank spaces and other non-alphanumeric characters in file names and subdirectory names will almost certainly cause problems. The compiled class files that are necessary for an applet to run will reflect these organizational principles.

After the code is written, it is compiled into an intermediate state called a class file. These class files contain the byte code for the Java virtual machine and have the "class" file extension. A typical programming project may produce dozens of class files. Since most Java programmers organize their projects into packages, there will often be a number of subdirectories, each containing one or more class files. When using an applet, you may copy the entire file structure containing an applet's class files and all the associated subdirectories to a new location on a hard disk, but you must not change the names of any files or subdirectories. Since Java has its roots in Unix, even capitalization can be important!

Although applets can be distributed as a collection of individual class files organized into subdirectories, there is a better way. Browsers are now capable of accessing and downloading a single "zip" file that contains everything necessary to run an applet. Java 1.1 takes this concept one step further and specifies a new file format called a Java Archive or "jar" file. Jar files are zip files on steroids; in addition to an applet's class files, a Java archive contains a manifest listing the archive's contents and encrypted security information such as the applet vendor. In the event of a "Class not found" error message, it is useful to open a jar file with a standard file decompression utility, such as WinZip, and see if the requested file is actually available. Class files are, of course, also distributed as part of a browser package, and it is possible that a "Class not found" error is due to a browser vendor that does not support the necessary functionality.

An applet begins its life cycle inside an HTML aware application after the class files have been downloaded. Since the Java machine does not really exist, Java byte code is interpreted one instruction at a time and converted into one or more native machine language instructions. These instructions are specific to the Pentium, Power PC, or whatever microprocessor is running the applet. This process is slow; it is the penalty to be paid for processor independence. Early Java Virtual Machines, Java VMs, performed this translation for every line of code as it was executed. Fortunately, browser vendors have now developed compilers that translate the entire class file into native machine code after downloading. These Just-In-Time, JIT, compilers have the potential of making Java almost as fast as C++ code. The shipping versions of Netscape Navigator and Microsoft Internet Explorer include such compilers.

Embedding

Insertion tags in an HTML document specify the type and location of multimedia content. For example, the <IMG> tag can be used to insert a 300 by 250 pixel image of an apparatus into a document using the following syntax:

```
<IMG SRC=http://physics.davidson.edu/images/apparatus.gif
  WIDTH="300" HEIGHT="250">.
```

Good WYSIWYG HTML editors are akin to the best word processors and hide these details from the author. The author simply does routine editing using cut and paste in order to insert images or highlighting to apply formatting and font styles. The finished document can be published on the Web. Unfortunately, high-level integration of advanced interactive Web-based technologies, such as JavaScript and Java, is still sketchy in some authoring packages. A passing knowledge of HTML and JavaScript syntax is usually required to develop interactive curricular material and for those times when it is absolutely necessary to get under the hood to find out what is going on.

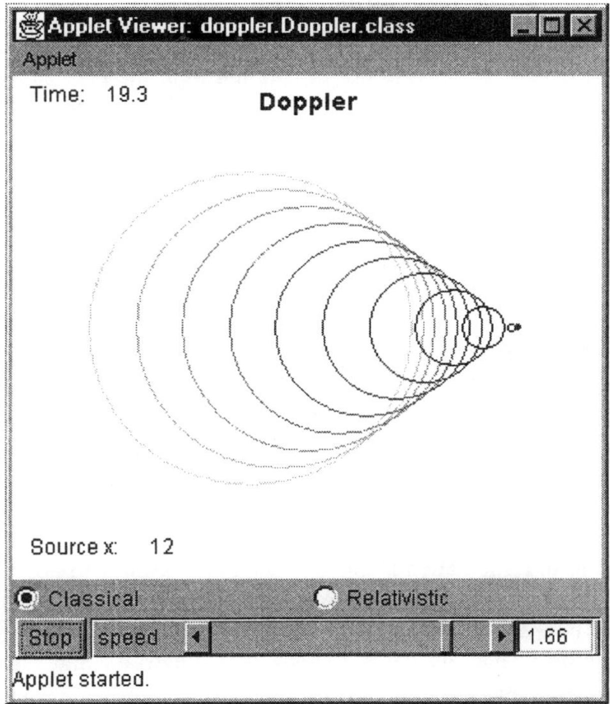

Figure 11.1 The Doppler Physlet as seen in the Sun Applet viewer.

Java 1.0 Embedding

Applets are embedded into an HTML document using the <APPLET> tag. This embedding is similar to adding a graphic to an HTML page with an <IMG> tag. Class files are downloaded into the browser along with other objects, such as sounds files or GIF images, which are referenced within the containing HTML document. It is now up to the browser to lay out the page on the screen and to translate the machine-independent class files into native binary code. It is the browser's job to provide the hooks into the operating system such as memory management. Many applets can be used effectively by merely embedding the applet into an HTML page and asking the student to describe the relevant physics. For example, the Doppler Physlet shown in Fig. 11.1 was used in the Giancoli Web site in problems 8 and 9 of Chapter 12. The applet can also be downloaded from Davidson College and embedded within a Web page using the following tags:

```
<APPLET CODE="doppler.Doppler.class" CODEBASE="classes"
  WIDTH=320 HEIGHT=370> </APPLET>
```

The browser displays the applet, and the user interacts with the applet using the applet's intrinsic controls. Doppler has a slider to set the velocity and an optional radio button to enable relativistic effects. In addition, the user can click-drag the mouse to make position measurements on the wave fronts. The input burden is on the user. The HTML author assumes the user is knowledgeable in the operation of the applet. This approach is easy and effective for simple applets. The CODEBASE attribute specifies that the browser should begin its search for the files necessary to run the applet in a subdirectory named "classes" relative to the directory containing the HTML document. Relative addressing using the usual dot-dot notation to move up one level may also be used to navigate the directory structure. For example, CODEBASE="../classes" would specify that class files are to be found by moving to the parent of the current directory and down into a subdirectory named classes. Just as with the <IMG> tag, you may specify a complete URL for the CODEBASE enabling you to access a Physlet—or any applet, for that matter—on an entirely different Web server from the one containing the HTML document. Due to bandwidth limitations and sluggish Internet response time, it is generally advisable to download and mirror an applet on a local file server as described in a later section.

Notice the CODE attribute inside the APPLET tag. The byte code for the applet is to be found in a file called Doppler.class, but this file will not be found in the subdirectory specified by the CODEBASE tag. Since the Doppler.class file is part of the Doppler package, it will be located in the doppler subdirectory. Appending the package name to the file name derives the CODE attribute. It is a common practice among Java programmers to use the same name for the package and the applet but without capitalizing the first character. Close examination of the downloaded files reveals that additional packages are required for this applet to function. The Graphics Java Toolkit is to be found in the gjt package located in a subdirectory of the same name (see *Graphic Java: Mastering the AWT* by David Geary, Prentice Hall 1998, for a complete description of this fine platform-independent package). The Doppler Physlet also uses other packages. These packages will be reused in other Physlets described below, but since most Physlets use the same supporting packages as Doppler, only a few new subdirectories containing new class files must be copied to the CODEBASE subdirectory.

A moderately complex Physlet may be prone to user error and frustration if too many parameters are presented on screen. Most authors prefer to add parameter (<PARAM>) fields to the <APPLET> tag in order to define the applet's behavior. Each applet has a unique set of parameter fields determined by the Java programmer. The programmer should, of course, document them. Applets can now be embedded multiple times with different conditions in order to solve different physics problems. The Superposition Physlet shown in Fig. 11.2 is used numerous times in

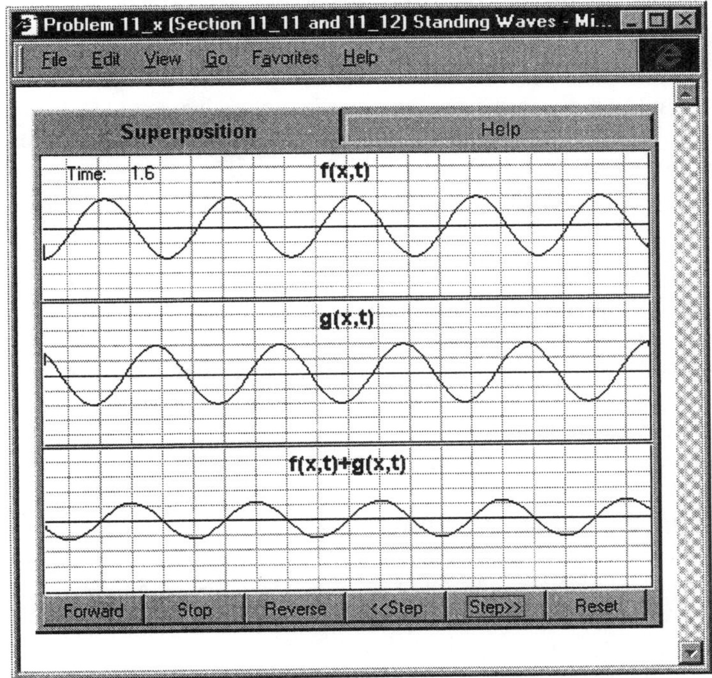

Figure 11.2 The Superposition Physlet displays two functions and their sum.

Chapter 11 of the Giancoli Web site. It uses two parameter fields to set arbitrary functions of position and time, f(x, t) and g(x, t). These functions are each displayed within a panel in the applet, and their sum is displayed in a third panel. This Physlet is set up to display a standing wave in an HTML page with the following tags:

```
<APPLET CODE="superposition.Superposition.class"
  CODEBASE="../classes/"
  WIDTH="404" HEIGHT="300">
      <PARAM NAME="numPoints" VALUE="100">
      <PARAM NAME="numGraphs" VALUE="3">
      <PARAM NAME="pixPerX" VALUE="20">
      <PARAM NAME="pixPerY" VALUE="10">
      <PARAM NAME="gridX" VALUE="1">
      <PARAM NAME="gridY" VALUE="1">
      <PARAM NAME="func1" VALUE="2*sin(pi*x/2-pi*t)">
      <PARAM NAME="func2" VALUE="2*sin(pi*x/2+pi*t)">
      <PARAM NAME="showControls" VALUE="true">
```

```
        <PARAM NAME="FPS" VALUE="10">
        <PARAM NAME="dt" VALUE="0.1">
</APPLET>
```

The parameter fields can easily be modified to produce other wave phenomena such as beats. A good way to learn about the superposition Physlet is to change a parameter and reload the HTML page to observe its effect. Superposition parameter fields are very extensive and allow a high degree of customization, as shown in the list below.

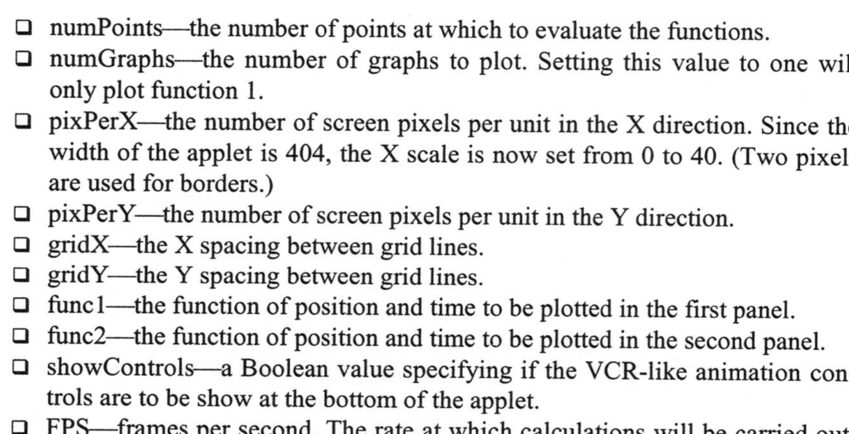

- ❑ numPoints—the number of points at which to evaluate the functions.
- ❑ numGraphs—the number of graphs to plot. Setting this value to one will only plot function 1.
- ❑ pixPerX—the number of screen pixels per unit in the X direction. Since the width of the applet is 404, the X scale is now set from 0 to 40. (Two pixels are used for borders.)
- ❑ pixPerY—the number of screen pixels per unit in the Y direction.
- ❑ gridX—the X spacing between grid lines.
- ❑ gridY—the Y spacing between grid lines.
- ❑ func1—the function of position and time to be plotted in the first panel.
- ❑ func2—the function of position and time to be plotted in the second panel.
- ❑ showControls—a Boolean value specifying if the VCR-like animation controls are to be show at the bottom of the applet.
- ❑ FPS—frames per second. The rate at which calculations will be carried out. Each frame requires that the functions func1 and func2 be evaluated at the specified number of points and then plotted on the screen. Too high a value and the computer may not be able to keep up. Too low a value and the user experiences screen flicker.
- ❑ dt—the simulation time increment for each frame. The simulation time has no relationship to the actual time.

Most applets will provide default values for parameters so that it is often not necessary to assign each and every parameter-name pair.

Although parameter fields spare the user from having to worry about many of the more arcane details of an applet, changing its behavior once it is embedded can still be far from foolproof. Parameter fields are read only once when the HTML page is loaded. The behavior of the applet will not change without user intervention. As an extreme example, consider the QTime Physlet used to display the time evolution of a quantum wave packet in an arbitrary potential, U(x). This applet has a tabbed panel that allows a user to access (and change) three function strings that define the real and imaginary parts of the wave function and the potential. This may be desirable for advanced students, but it is unlikely that a sophomore-level modern physics student needs to be confronted with such detail. Few students would know how to write the real and imaginary parts of a Gaussian wave function in atomic units with the appropriate momentum boost needed to produce a reasonable group velocity. Almost all Physlets have a "showControls" parameter that is designed to simplify the user interface and to hide the user interface controls. Changing parameters, as well as starting and stopping the animation, must now be done using a scripting language as explained in the following section.

Java 1.1 Embedding

Both applets discussed so far were written in Java 1.0 and are distributed as uncompressed class files in order to maintain maximum compatibility with early browsers. Newer applets are distributed using the Java Archive, jar, format as described in the previous section. These applets require that the ARCHIVE attribute be added to the <APPLET> tag. The Optics applet used throughout chapters 23 and 25 of the Companion Web site is embedded as follows:

```
<APPLET CODEBASE="../classes/" ARCHIVE="Optics.jar"
  CODE="optics.OpticsApplet.class"
  WIDTH="400" HEIGHT="200">
      <PARAM NAME="ShowControls" VALUE="true">
      <PARAM NAME="PixPerUnit" VALUE="100">
</APPLET>
```

The CODE attribute's format alerts us to the fact that there will be an optics subdirectory inside the archive if it is unpacked.

Java 1.1 and JavaScript to Java communication are in the processes of being implemented on all major platforms. Unfortunately, support on current browsers is still spotty. Netscape Communicator, version 4.05 and below, does not support Java 1.1 and will not run these applets. Microsoft Internet Explorer 4.0 does not support JavaScript to Java communication on the Apple Macintosh although the Optics Physlet will run in interactive mode, i.e., without script, using the latest version of the Apple Java VM. Java 1.1 Physlets seem to run best on Internet Explorer 4.0 using Windows 95/NT and on browsers using the Sun Java VM plug-in.

JavaScript to Java Communication

Java and JavaScript have little in common but are excellent companions. Java is similar to C in some respects, but its philosophy is much closer to Object Pascal or the object-oriented language Smalltalk. JavaScript is an interpreted scripting language that is closer to Visual Basic. Java has the industrial strength to provide the computational bricks to build robust interactive problems while JavaScript is the mortar that ties these bricks together. In order to JavaScript an applet, an additional attribute must be added to the <APPLET> tag in order for the JavaScript interpreter to be able to identify the applet.

```
<APPLET CODEBASE="../classes/"CODE="Hello.class"
  WIDTH="400" HEIGHT="200" NAME="Hello" ID="Hello">
  </APPLET>
```

Netscape uses "Name" and Microsoft uses "ID" in order to refer to the embedded applet, so it is a good idea to define both attributes using the same string value. The strings attached to these two identifiers need not be the name of the class file, although they often are if the applet will only appear once on a page. Use different names if you wish to embed any applet more than once in a page so that each applet can be referred to by its unique name.

The following code shows a very simple applet that can be embedded and scripted from within an HTML page to display a text message. The paint(Graphics g) method draws a string near the center of the applet. It is invoked whenever the browser determines that the applet needs to be drawn. The applet also implements two methods to set and get the displayed message string.

```
import java.applet.*;
import java.awt.*;
public class Hello extends Applet
{
    private String message="Hello World";
    public void init(){
           this.setBackground(Color.white);
    }
    public void setMessage(String m){
           message=m;
           repaint();
    }
    public String getMessage(){
           return message;
    }
    public void paint(Graphics g){// paint the string
           Rectangle r=this.bounds();
           g.drawString(message, r.width/2, r.height/2);
    }
}
```

The visibility of variables and methods is controlled through the public and private keywords. For example, the variable message is declared private and is not accessible from JavaScript (or from other Java classes). Access to this variable is controlled by the accessory methods getMessage() and setMessage(String m). These methods can be invoked from JavaScript without any additional formalities using standard JavaScript syntax:

```
document.hello.setMessage("Another Message").
```

Any public methods in a Java applet can be accessed using JavaScript. These methods are, however, applet specific, and you will need to consult the applet's documentation to determine proper calling signatures. A sample script is shown below.

The most efficient way to include JavaScript in an HTML page is to define the script in the <HEAD> section of an HTML page. A typical head would contain the following tags:

```
<HEAD>
<TITLE>Describing Motion</TITLE>
<SCRIPT language="JavaScript">
function prob1(){
      funcX="5*t*t";
      funcY="0";
   document.Animator.setDefault();
      document.Animator.shiftPixOrigin(-150,0)
      document.Animator.setShapeCoord(4);
      document.Animator.setShapeRGB(255,0,0);
      document.Animator.addRectangle(50,
         15,funcX,funcY);
      document..setShapeCoord(0);
      document.Animator.setShapeRGB(0,0,0);
      document.Animator.addArrow("5","0",funcX,funcY);
      document.Animator.setTimeInterval(0,3);
      document.Animator.forward();
}
</SCRIPT>
</HEAD>
```

This script is based on the Animator Physlet. It is used with Problem 1 in Chapter 2 of the Giancoli Web site. It draws a red rectangle and a black arrow. These objects move in the x and y directions according to the analytic functions specified by the string variables funcX and funcY, respectively. It is important to note that some parameters are passed as integer values while other parameters are passed as strings. String values usually denote a function that will be evaluated repeatedly while the animation takes place. Integer values, on the other hand, usually denote fixed values that cannot be changed once the object has been created. A complete description of the public methods in Animator Physlet can be found in a subsequent section (Animator Documentation). Documentation for other Physlets can be found on the Davidson WebPhysics site.

There are a number of ways to invoke a JavaScript function from within the body of an HTML page. The easiest method is to use a variant of the anchor tag. The anchor tag, <A>, is most commonly used to link to another HTML page

```
<A HREF="http://webphysics.davidson.edu">Go to WebPhys-
   ics!</A>
```

but it can also be used to send e-mail, connect to an ftp server, or execute JavaScript. In order to execute a JavaScript function, the http attribute is replaced with the JavaScript attribute as follows:

```
<A HREF="JavaScript:prob1()">Initialize problem 1.</A>
```

Clicking on the text bracketed by the <A> and </A> tags will execute the JavaScript function that was previously defined as prob1(). This technique makes it possible to use a single Physlet to do many different physics problems and to change its behavior while the student is working through the exercise.

In summary, a complete HTML page containing an interactive Physlet problem will usually contain the following elements: (1) a <HEAD> containing JavaScript functions capable of communicating with the Physlet, (2) a body containing a Physlet embedded with the <APPLET> tag, and (3) an anchor tag to invoke the appropriate JavaScript. A complete HTML page is shown below for reference.

```
<!DOCTYPE HTML PUBLIC "-//IETF//DTD HTML//EN">
<html>
<head>
<meta name="GENERATOR" content="Microsoft FrontPage 3.0">
<title>Describing Motion</title>
<script language="JavaScript">
function prob1()
{       document.Animator.setDefault();
        document.Animator.shiftPixOrigin(-150,0)
        document.Animator.setShapeCoord(4);
        document.Animator.setShapeRGB(255,0,0);
        document.Animator.addRectangle(50,
           15,"5*t*t","0");
        document.Animator.setShapeCoord(0);
        document.Animator.setShapeRGB(0,0,0);
        document.Animator.addArrow("5","0","5*t*t","0");
        document.Animator.setTimeInterval(0,3);
        document.Animator.forward();
}
</script>
</head>
<body bgcolor="#FFEFCE">
<p align="center">
<applet code="animator.Animator.class" code-
   base="../classes/" align="baseline"
   width="400"height="150" id="Animator" name="Animator">
        <param name="dt" value="0.02">
        <param name="FPS" value="20">
        <param name="gridUnit" value="1.0">
        <param name="pixPerUnit" value="10">
        <param name="showControls" value="false">
</applet>
</p>
```

```
<h2>Describing Motion</h2>
<p><i>Be sure the simulation has finished loading before
   you click begin.</i></p>
<p>Physicists use arrows to represent many things in dia-
   grams. What vector quantity is being represented by
   the arrow in this simulation? 
<a href="JavaScript:prob1()">Start simula-
   tion</a> </p>

<p><input type="radio" name="answer.1" value="1"> dis-
   placement</p>
<p><input type="radio" name="answer.1" value="1"> veloc-
   ity</p>
<p><input type="radio" name="answer.1" value="1"> accel-
   eration</p>
<p><input type="radio" name="answer.1" value="1">
   speed</p>
<p><strong>Instructions</strong></p>
<p>Click on start simulation to initialize the simulation
   at the top of the page. Then answer the question after
   you have run the simulation.</p>
</body>
</html>
```

Fig. 11.3 shows a screen shot of this HTML code after it has been incorporated into the Giancoli Companion Web site.

Animator Scripts

The Animator Physlet is a scriptable drawing applet designed to display and animate free-body diagrams and particle trajectories. The animation is along a prescribed trajectory and need not obey any physical law. This makes Animator ideal for testing student misconceptions since both physical and nonphysical motion can be scripted. This section presents three such Animator scripts.

A kinematics example

A common misconception among students is confusion between a quantity and its rate of change. The Physlet problem illustrated in Fig. 11.4 tests for this misconception by presenting two animations with identical accelerations. However, in the first animation the object is moving up, and in the second the object is moving down. Two functions are defined in the head and called from HTML anchors as described previously. The script is shown below.

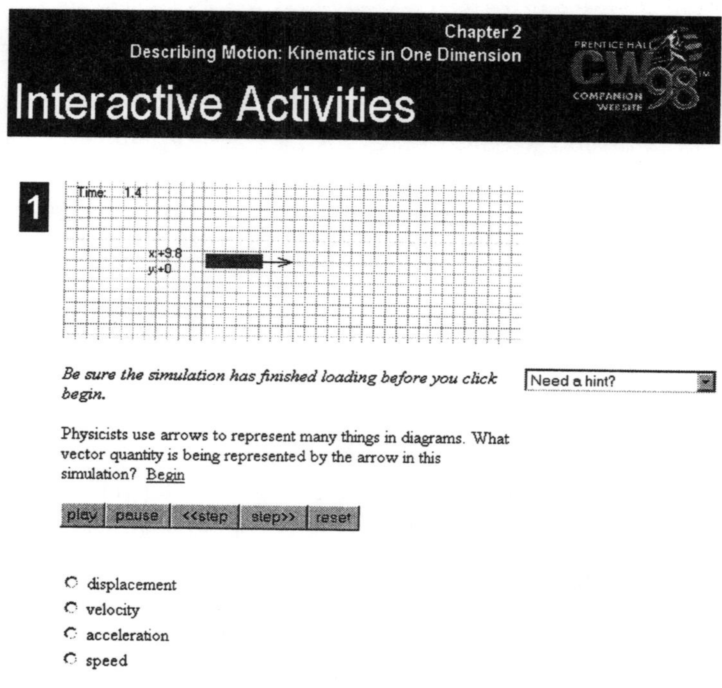

Figure 11.3 A Physlet problem based on the Animator Physlet, as it appears on the Giancoli Companion Web site.

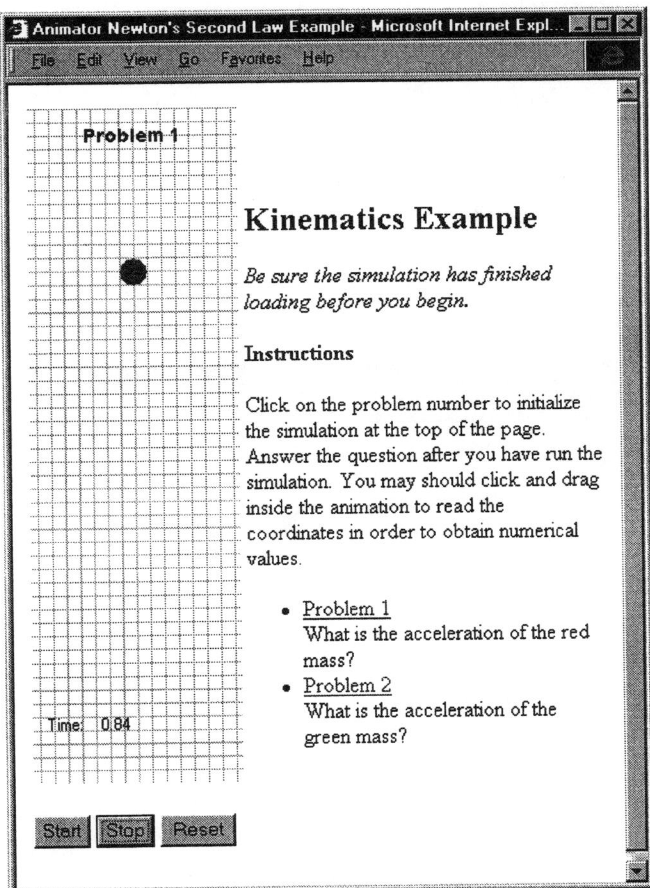

Figure 11.4 A kinematics Physlet problem based on the Animator Physlet.

```
<script language="JavaScript">
function prob1(){
     document.Animator.setDefault ();
        document.Animator.setShapeRGB(255,0,0);
        document.Animator.addCircle(20,"0","20-10*t*t");
        document.Animator.setCaption("Problem 1");
        document.Animator.setTimeInterval(0,2);
        document.Animator.forward();
}
```

```
function prob2()
{       document.Animator.setDefault ();
        document.Animator.setShapeRGB(0,255,0);
        document.Animator.addCircle(20,"0","-25+(45*t)-
           10*t*t");
        document.Animator.setCaption("Problem 2");
        document.Animator.setTimeInterval(0,2);
        document.Animator.forward();
}
</script>
```

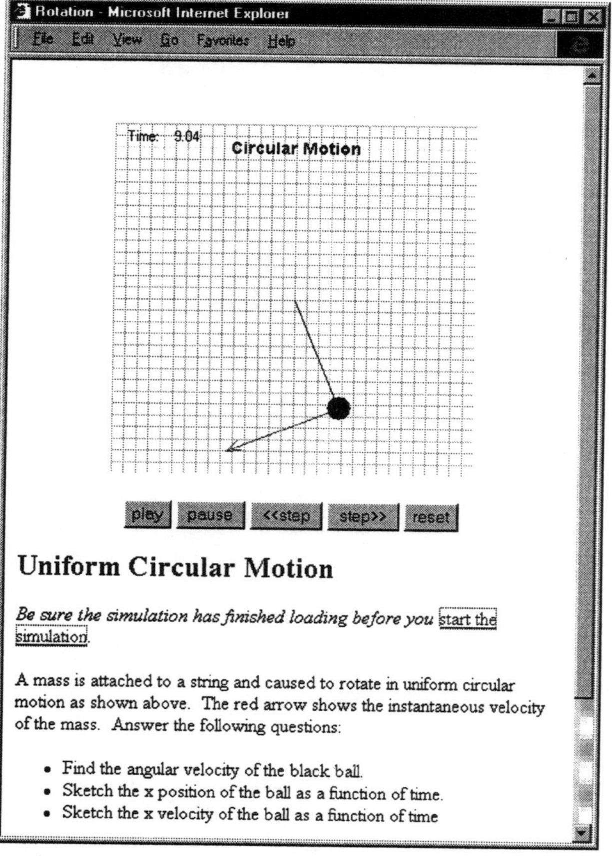

Figure 11.5 A simulation of uniform rotation as seen on Windows 95 with Internet Explorer.

Most Animator scripts begin with a setDefault() method in order to stop the animation, clear all previous drawing, and put the applet into a predefined state. The second statement in each function sets the red, green, and blue color components of the fill color for all subsequent objects. These components must be in the range [0...255]. The third statement is the most important. It creates a circle that will move inside the animator along the scripted trajectory. The first parameter is an integer value that specifies the circle radius in pixel units. The following two parameters are string variables that specify the x and y coordinates of the circle as functions of time. Notice that even though the x coordinate takes on the constant value of zero, it must still be input as a string. The formula for the y coordinate produces a constant acceleration of 20. The next to last statement specifies that the animation should loop over the time interval [0, 2]. The default behavior would have incremented the animation time forever, i.e., until a numeric overflow occurs. The forward() method in the last statement starts the animation.

In order to keep the user interface as simple as possible, the "showControls" parameter was set to false when the applet was embedded. The buttons that appear on the screen are standard HTML form elements. The action methods of these buttons invoke the appropriate methods to start and stop the animation on the screen.

```
<form>
<input type="button" value="Start" on-
   click="document.Animator.forward()">
<input type="button" value="Stop" on-
   click="document.Animator.stop()">
<input type="button" value="Reset" on-
   click="document.Animator.reset(0.0)">
</form>
```

The onclick actions for the three buttons are set equal to three different JavaScript statements. They could just as well call multi-line JavaScript functions. Using forms and buttons to call script rather than using anchor tags is purely a question of style.

Circular Motion

The connection between simple harmonic motion and circular motion is used in many non-calculus Physics texts to derive the relationships among velocity amplitude and frequency (see, for example, Giancoli Chapter 11 Section 4). A simple Animator-based exercise (illustrated in Fig. 11.5) can help students understand the basis for this derivation. The script for this exercise makes use of the addLine() and addArrow() methods.

```
<script language="JavaScript">
function prob1(){
      x="10*sin(t)";
      y="10*cos(t)";
```

```
        vx="10*cos(t)";
        vy="-10*sin(t)";
        document.Animator.setDefault();
        document.Animator.addLine(x,y,"0","0");
        document.Animator.addCircle(20,x,y);
        document.Animator.setShapeRGB(255,0,0);
        document.Animator.addArrow(vx,vy,x,y);
        document.Animator.setCaption("Circular Motion");
        document.Animator.forward();
}
</script>
```

This script begins by defining four string variables to hold the position and velocity components of the mass. These variables are then passed as parameters to addCircle(), addLine(), and addArrow(). The first two arguments to either method specify the displacement relative to the position of the object. The position of the line is fixed at the origin (0, 0), while the position of the velocity vector is set to the position of the circle.

Conservation of Momentum

Conservation of energy and momentum are two equally important physics principles, yet students seem to have an easier time applying the first conservation law than the second. Physlet animations showing various types of collisions make excellent JiTT exercises or concept questions. The following script can be used as a guide to writing other collision problems.

```
<SCRIPT LANGUAGE="JavaScript">
function prob() {
        x1= "step(0.55-t)*4*t+step(t-0.55)*(2.2-2*(t-
           0.55))";
        x2= "step(0.55-t)*(3.5-2*t)+step(t-
           0.55)*(2.4+1*(t-0.55))";
        document.Animator.setDefault();
        document.Animator.shiftPixOrigin(0,0);
        document.Animator.setPixPerUnit(100);
        document.Animatordocument.Animator.shiftPixOrigin(
        -175,0);
        document.Animator.setShapeCoord(4);
        document.Animator.setShapeRGB(0,0,255);
        document.Animator.addCircle(20,x1,"0");
        document.Animator.setShapeCoord(2);
        document.Animator.setShapeRGB(255,0,0);
        document.Animator.addCircle(20,x2,"0");
        document.Animator.forward();
}
```

```
</script>
```

The x position of the red and blue circles make use of the Heaviside unit step function, step(z), in order to change velocity at t = 0.55.

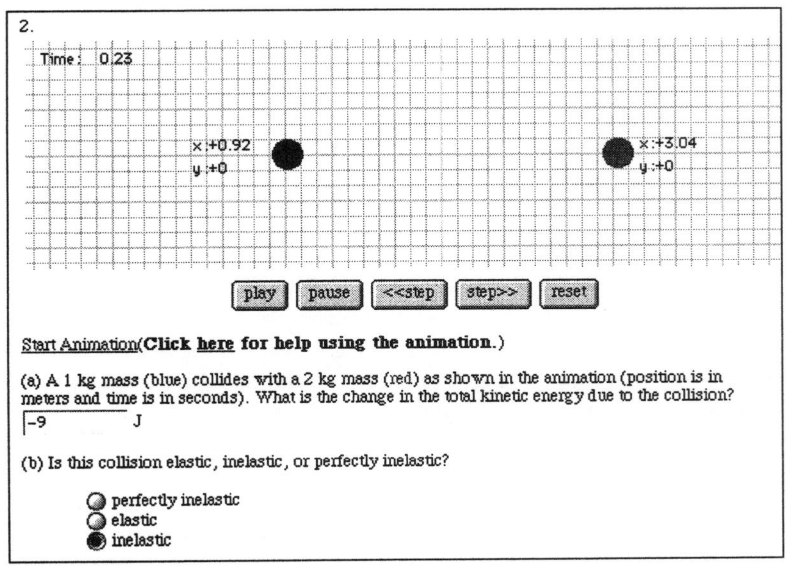

Figure 11.6 A collision problem as seen on a Macintosh using Netscape Navigator.

Available Physlets

There are currently 24 Physlets available on the Davidson College WebPhysics server,

http://webphysics.davidson.edu/applets/applets.html.

The following nine Physlets were developed for the Prentice Hall Web site and are the most suitable for introductory physics.

Java 1.0 Physlets

- **Animator:** A general-purpose drawing applet used to animate trajectories. Available shapes include circles, rectangles, polygons, lines, and arrows. The trajectory that each shape follows is specified when the shape is constructed. Used extensively in mechanics problems. See Giancoli Companion Web site Chapters 2–10.

- **Doppler:** A simple but useful Doppler-effect applet. It has become our most downloaded applet. A radio button allows the user to compare classical and relativistic Doppler shifts.
- **Superposition:** Designed to display the sum of two traveling waves thereby demonstrating wave superposition effects such as standing waves and beats. This applet can also be used effectively to show the relationship between frequency, wavelength, and speed by restricting the display to a single animated function.
- **Ripple Tank:** A realistic simulation of point sources in a water-filled ripple tank. Useful for displaying multiple-source interference patterns and for showing the progression of interference patterns as the number of sources is increased. Sources can be rearranged by click-dragging if the applet is placed into edit mode.

Java 1.1 Physlets

- **Bfield:** Display the magnetic field in the x-y plane. B_x and B_y may be scripted as well as any number of current carrying wires. Two wire geometries are supported: (1) long, straight wires in the z direction and (2) circular loops of arbitrary radius whose axis lies parallel to the x axis.
- **EField:** Display electric field lines, direction fields, and equipotential lines for an arbitrary collection of point charges. An external potential function, U(x, y), may also be scripted. In addition, test particles may be placed inside the field and their subsequent motion observed.
- **Circuits:** A collection of applets for the study of ac and dc circuits, including RC and LRC circuits.
- **Faraday:** The classic gedanken experiment consisting of a U-shaped fixed wire and a slideable straight wire on the open side. Click-drag the slide wire or script a time-varying magnetic field and observe the resulting *emf* on the meter and on the time graph.
- **Optics:** A general-purpose optics bench capable of displaying multiple optic elements, including mirrors, lenses, light beams, and point sources. Properties such as position and focal length may be adjusted in real time by click-dragging the optic element.

Downloading

Obtaining the latest version of a Physlet requires a browser, an Internet connection, and a file decompression program capable of extracting files from a ZIP archive. Point any standard browser at the Davidson WebPhysics site,

http://webphysics.davidson.edu/applets/applets.html

and click on the link to the download page. This page contains a list of anchor tags to the various Physlet archives. Clicking on a link should bring up a dialog box asking if you want to open or save the corresponding archive. Choose the save option and save the file to a temporary directory on your hard drive. Open the file after it has downloaded and decompress the entire archive contents into a working directory. The working directory will contain a small HTML file that should run the Physlet as is. Open this HTML file with a Java aware application and check to make sure that it runs. A nonrunning applet is most likely caused by not having a Java 1.1 compliant browser. Move the class files to a permanent directory and adjust the codebase in the HTML file to refer to this directory. Happy scripting!

Animator Documentation

Documentation is available on each Physlets home page. Animator documentation is provided here for use in workshops and other off-line environments. Animator may be scripted using the following Java method calls:

- **addArrow(String hStr, String vStr, String xStr, String yStr)**
  An arrow, i.e., a vector, that moves according to the functions x(t)=xStr and y(t)=yStr. The horizontal, hStr, and vertical, vStr, components can be functions of time to allow for rotating vectors.
- **addCircle(int size, String xStr, String yStr)**
  A circle that moves according to the functions x(t)=xStr and y(t)=yStr. The size s is in pixels. The method replaces the addShape method in previous versions.
- **addLine(String hStr, String vStr, String xStr, String yStr)**
  A line segment that moves according to the functions x(t)=xStr and y(t)=yStr. Behaves similarly to arrow. The horizontal, hStr, and vertical, vStr, components can be functions of time to allow for rotating lines segments.
- **addParametricCurve(int n, double start, double stop, String xStr, String yStr)**
  Add a parametric curve defined by the functions x(t)=xStr and y(t)=yStr. N points will be drawn starting at s=start and ending at s=end.
- **addPolyShape(int n, String hStr, String vStr, String xStr, String yStr)**
  Add an arbitrary shape having n vertex points. Can be used to add moving triangles, rotated rectangles and other arbitrary shapes to the animation. **Notice:** The x and y vertex points are passed to the Physlet as String variables (i.e., not arrays) since this data type is common to Netscape and Internet Explorer. The Shape will move according to the functions x(t)=xStr and y(t)=yStr.
- **addRectangle(int w, int h, String xStr, String yStr)**
  A rectangular shape that moves according to the functions x(t)=xStr and y(t)=yStr. The width, w, and height, h, are in pixels.
- **addText(String text, String xStr, String yStr)**
  Text that moves according to the functions x(t)=xStr and y(t)=yStr. Can be used to label moving shapes since the text is drawn last.
- **deleteAll()**
  Remove all rectangles, arrows, and circles from the animator.

- **forward()**
  Run animation with positive time step.
- **pause()**
  Pause the animation.
- **reset(double time)**
  Set time.
  reverse()
  Run animation with negative time step.
- **setCaption(String caption)**
- **setInterval(double start, double stop)**
  Time will reset to the start value whenever time exceeds the stop value.
- **setShapeCoord(int val)**
  Display the coordinates of a circle, rectangle, or vector. Call this method *before* you add the object. All objects added will display their coordinates according to the following: val=0, not coordinate display; val=1, coordinate display above the object; val=2, coordinate display to the right of the object; val=3, coordinate display below the object; val=4, coordinate display to the left of the object.
- **setShapeCoord(int val)**
  Display the coordinates of a circle, rectangle, or vector. Call this method *before* you add the object. All objects added will display their coordinates according to the following: val=0, not coordinate display; val=1, coordinate display above the object; val=2, coordinate display to the right of the object; val=3, coordinate display below the object; val=4, coordinate display to the left of the object.
- **setShapeRGB(int r, int g, int b)**
  Sets the fill color for new shapes.
- **setShapeTrail(int n)**
  Sets the trail length for new shapes. May cause unpredictable results due to a change in syntax between Java 1.02 and Java 1.1.
- **stepBack()**
  Step the animation back by negative dt.
- **stepForward()**
  Step the animation forward by negative dt.
- **start()**
  Start animation thread. It is usually better for the JavaScript programmer to call forward()or reverse() and let the browser manage the thread.
- **stop()**
  Stop the animation thread. It is usually better for the JavaScript programmer to call pause() and let the browser manage the thread.
- **shiftPixOrigin(int xo, int yo)**
  Shift the origin, i.e., x = 0 and y = 0 point, away from the center of the applet. Example: shiftPixOrigin(–100, 30) will shift the origin 100 pixels to the left and 30 pixels up.

Chapter 12: Communication

by Larry Martin, North Park University
and
Aaron Titus, North Carolina Agricultural and Technical State University

Successful use of JiTT requires rapid, asynchronous, electronic communication between teachers and students. A key aspect of JiTT is that students must be actively engaged with the material rather than being passive recipients. E-mail can be sent directly to and from students, while Web pages are typically open to the world (although methods exist for limiting distribution to a set of users or locations). Further tuning of the communication process can be done with the use of CGIs (Common Gateway Interfaces) to control content and feedback automatically. The communication methods described in this section should not be viewed as a hierarchy of effectiveness, but rather as a list in order of increasing complexity during initiation resulting in some efficiency during later stages.

E-mail

One of the simplest, most familiar, and readily available electronic communication technologies is e-mail. The asynchronous nature of e-mail is as natural to students and faculty as phone answering machines. Distributing and collecting WarmUps, Puzzles, and other assignments by e-mail is easy and effective. E-mail clients with powerful GUIs and sorting tools can make it easy to keep track of responses. A constant stream of student answers to homework might be overwhelming for even the most assiduous teacher, but immediate availability and familiarity makes it worthwhile.

Further enhancement of the e-mail learning process can be made by "publishing" responses, either by forwarding some or all of the answers to the class, or by arranging a list-serve to forward all responses to the entire class. This has the advantage of bringing peer pressure to bear on participation and can increase the sense of responsibility to "get it right." We have found that writing also improves when it is not "just the teacher" who is reading it. But this quickly becomes too burdensome when the traffic becomes high. Students can easily be overwhelmed by the amount of e-mail generated even by a small class of ten students. The large volume of responses going out to everyone also makes it harder for the late joiners to feel they have something to contribute. The effect is to quickly squelch participation once a few adequate responses have been made. Early arrivals tend to dominate list-serve discussions. E-mail is a successful tool for JiTT when used to create a sense of ownership of ideas on the part of the students. When discussing e-mail in class, it is encouraging to cite the source or ask the contributor to repeat the essence of the idea. This can mitigate the feelings of neglect if teachers cannot reply to every e-mail generated by students.

E-mail may be easily archived onto the Web with tools like

> Lyris: http://www.lyris.com/,
> HyperMail: http://www.landfield.com/hypermail/ or
> MhonArc: http://www.oac.uci.edu/indiv/ehood/mhonarc.html.

This has the advantage of placing the e-mail into an accessible location while not stuffing everyone's in-box with all the e-mail from the class. The Web-based e-mail is automatically sorted by date, topic, or author, allowing quick evidence of participation and quick retrieval of items for class discussion.

HTML Forms

Electronic distribution of more complex information is easier in HTML than in e-mail—pictures and Physlets may be deployed rapidly and simply on the Web. HTML alone, however, can encourage passivity. HTML forms require students to engage with the material. Forms provide a means by which students can be asked to interact with the material by providing a response to one or more questions. It is even possible for a student's responses to be analyzed, and a new page returned providing feedback or further questions (see the section on CGIs, below). Forms can contain multiple choice, fill-in-the-blank, and essay-type responses to questions. Their inclusion in Web pages is reasonably easy, although some WYSIWYG HTML editors do not do as good a job with forms as they do with other elements. An example of a form is a page from a search engine or Internet shopping market. Users type information into the visible representation of the form on the screen, i.e., the text boxes. But nothing happens to this information until it is sent to another program. A form can be submitted by e-mailing the results to the teacher, or it may be processed by a CGI on the server.

Forms allow (and teachers can require) submission of answers to carefully chosen questions in the midst of the information being presented. Forms containing JavaScript and/or Physlets can be constructed to automate the process and give immediate feedback to students [Titus et al., 1998]. In JiTT, students are highly motivated to use the forms, because they know that their responses, e.g., to WarmUps, will be used in classroom discussions.

As an example of a simple form and an action based on the form, the following code presents a WarmUp question for discussion (as shown in Fig. 12.1), and directs the student's response to the teacher's e-mail. The form elements where the student enters a name and an answer are created with the <INPUT> and <TEXTAREA> tags, respectively. The <FORM> tag is used to specify the action that will be performed when the form is submitted. In this case, the action is mailto:teacher@myplace.edu. During the class, a discussion is held on the relative merits of a few selected answers posted before class (Fig. 12.2).

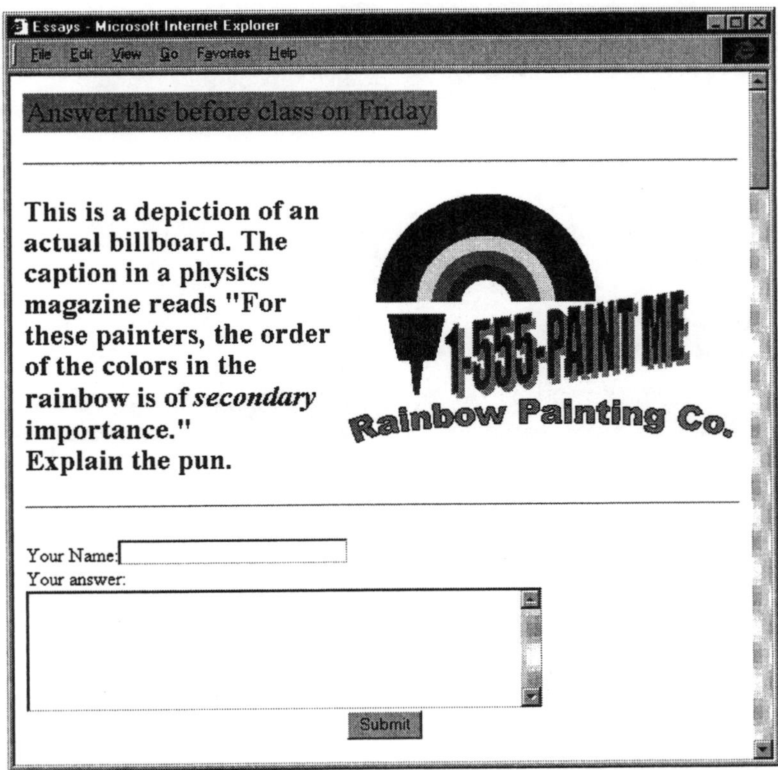

Figure 12.1 The page generated by the HTML in the previous example.

```
<HTML><HEAD><TITLE>Essays</TITLE></HEAD><BODY>
<TABLE BGCOLOR='#00DFDF'><TR><TD>
<FONT SIZE=+2 COLOR="red">Answer this before class on
   Friday</FONT>
</TD></TR></TABLE>
<P><HR><P>
<IMG SRC="rainbow.gif" align=right>
<H2>This is a depiction of an actual billboard.
The caption in a physics magazine reads
```

```
"For these painters, the order of the colors in the rain-
   bow is of
<I>secondary</I> importance."
<BR>Explain the pun.</H2>
<P><HR><P>
<FORM METHOD="POST"
ACTION=mailto:teacher@myplace.edu ENCTYPE="text/plain">
Your Name:<INPUT TYPE="text" NAME="Students" VALUE=""
   SIZE=25 >
<BR>Your answer:
<BR><TEXTAREA NAME="Essay" ROWS=6 COLS=50
   WRAP=SOFT></TEXTAREA>
<BR><CENTER><INPUT TYPE="submit" NAME="action"
   VALUE="Submit"></CENTER>
</FORM>
</BODY></HTML>
```

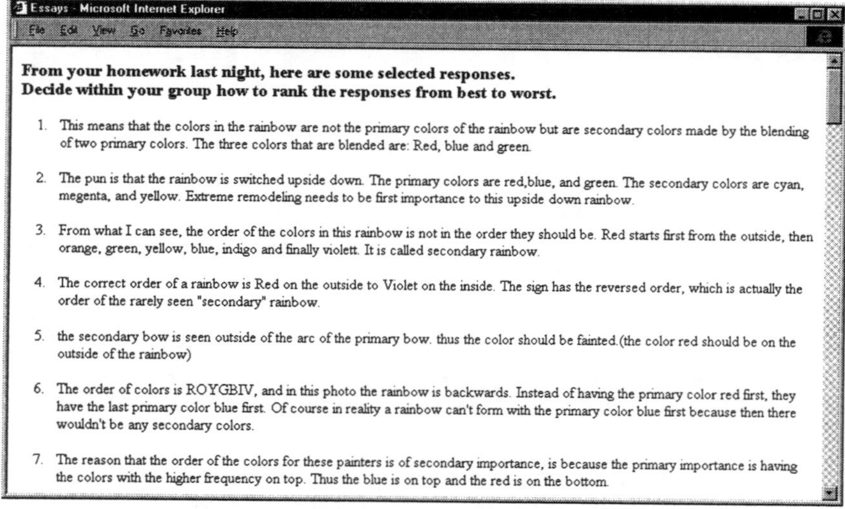

Figure 12.2 A few responses to the question shown in Fig. 12.1, as they might be presented for a class discussion.

CGI

CGI (Common Gateway Interface) applications are programs that are run on the Web server and can handle several tasks, including creating Web pages "on the fly," recording and responding to submissions from HTML forms, and interacting with other resources such as databases or measurement/control peripherals. Since almost any program can be run as a CGI application, including programs that are part of the operating system, the CGI standard enforces stringent security to protect the Web server. For example, CGI programs must be placed in special directories on the Web server so, it is impossible to run programs such as "rm *" or "FORMAT C:" unless the server administrator has foolishly copied these programs into these special directories.

For JiTT, CGIs are best used to ease the burden of collecting, recording, and confirming student submissions. Having a CGI respond immediately to the student decreases the student's desire for an immediate response to e-mail. A CGI can also handle the task of formatting simple questions as HTML forms. This frees the teacher to spend time on devising interesting activities and thoughtfully engaging the students where they have difficulty.

CGIs can be written in any computer language that will run on the platform on which the Web server is running. C++, Perl, Basic, HyperCard, and ToolBook have all been used in creating CGIs. Although Perl continues to be the most popular language for writing CGIs, some new programming tools, like Tango, have been created specifically for writing CGIs using a graphical interface. The use of a CGI brings a new immediacy to the idea of JiTT—Web pages can be generated according to the needs of the student in the millisecond they are needed!

The most common way that we have used CGIs is in the collection of student responses to the WarmUps; the CGI collects the responses and delivers consolidated reports to the teacher. This may be as simple as a summary of students' written responses or as sophisticated as a statistical report of the spread of the students' responses, assisting the teacher in designing the next class or assignment. A number of other features may be implemented in CGIs, depending on the programming prowess of the teacher or staff and the needs of the students. Besides automatic grading of simple forms, CGIs can aid in delegating tasks among teams, redistributing responses for peer review or in-class analysis and discussion, and maintaining deadlines for all work. By having CGIs store the submissions for later review, the teacher is insulated from the barrage of e-mail that results from other techniques.

Another way in which a CGI can be used in JiTT is as follows: A Web page link leads a student to a CGI that creates a login page if the student has not been previously authenticated. The student fills in identifying information (perhaps authenticating with a password), and the Submit button then takes the student to a CGI, which creates an assignment tailored to that student. That CGI may look up questions selected from a database, modify the selection based on the student's history, and personalize some elements of the question. The student receives the page and responds to the questions, and on submitting the form, a CGI records the answers and creates a page of appropriate feedback. The feedback may simply confirm receipt or produce a report of correctness, perhaps even with the possibility of further submissions. If students are able to retry, there often are fewer misunderstandings remaining on basic information when the class next meets and the teacher may concentrate on the open-ended problems (which are not automatically graded), as well as on other questions that remain. It is advisable in the context of JiTT that the feedback be used more as motivation than as evaluation so that closure in class discussion may be reached without heightened anxiety about getting "points."

Important Considerations

There are several issues to be considered before implementing CGI-based JiTT. One issue is expense. Evaluate the total investment before committing large resources of equipment and personnel to developing a set of robust and useful CGIs. By far the most expensive portion is in personnel, either in institutional staff or the teacher's own time. Do not be fooled into thinking the equipment is the expensive part! Of course, one can always use or modify programs written by others. "WWWAssign," available from

http://www.northpark.edu/~martin/WWWAssign,

was one of the earliest tools for Web-based assignments; versions have been available for free downloading since 1995, with several hundred people using it worldwide. Several institutions have built additional resources into the base code and successfully used the program for their classes. We are currently writing a new version, WebAssign, which will be available in both free and commercial versions. The new program will include database connectivity. Using programs that others have written is one way to leverage existing applications and save on personnel costs. Surely no one has to write their own browser, now that Netscape and Microsoft exist. It is best to use browsers, servers, spreadsheets, or databases that are widely used in commerce and industry so that the software underlying your JiTT applications will not disappear in the near future.

Another issue is the server—personal desktop machine versus large-scale deployment. It is now fairly easy to deploy a Web-server running CGIs from a teacher's desktop machine. Many teachers routinely publish their Web pages in this fashion. Adding CGIs causes the load to increase substantially on such machines, and the machines must be able to handle the peak load, not simply the average load. Having a Web-based assignment due at 5 p.m. makes the machine essentially unusable for any other purpose during the last few minutes of that time. Large-scale deployment is typically done on a dedicated machine, usually shared among all teachers at an institution and managed by a staff who are motivated by concerns wider than what one teacher may want to do. Security is typically higher, which is both an advantage and a disadvantage. It is helpful for teachers not to have to manage their own set of usernames and passwords for their own students. Allowing CGIs to be freely changed may conflict with system administrators' needs to maintain a stable system for everyone. But such systems also have a large staff dedicated to keeping the system up 24 hours a day, 7 days a week, and backups and disaster-recovery plans are typically already in place. A desktop machine can easily go down without the teacher knowing about it until the next weekday. But the open availability of the personal desktop machine allows greater freedom to try things out without annoying other users.

Another issue is that of expertise. A large-scale shared system typically has many people invested in keeping it going. Expertise and help can be gained from the existing users and staff. Teachers need good institutional support for constant Internet access and domain name assignments. Many institutions run DHCP (Dynamic Host Configuration Protocol) for desktop machines, which means the machine cannot be found on the Internet easily since the IP address changes every time the machine is rebooted. A desktop system is typically regarded as "your problem," although the widespread commercial use of personal computers has created an expanding base of expertise that can be consulted in newsgroups and by e-mail. Installing and using a personal Web-server can take anywhere from a day to several weeks, depending on your existing knowledge. Installing and managing Web-servers on personal computers is relatively easy, and some good Web-server software is available free of charge.

Creating CGIs requires a significant depth of understanding of programming, often in several languages. In writing CGIs, it is helpful to test the outcome on several different platforms. By maintaining code that works for all browsers, students are isolated from the glitches that occur because of platform issues. "Write once, run everywhere" is a marketing slogan. It has nothing to do with reality.

A clear and maintainable plan is needed to organize the contents of any Web site. Several questions should be answered before launching a site. How will content be organized so that it can be found again next year, edited and reused, or moved easily from one machine to another as needs change? Who will use and maintain the pages and the CGIs? Who will pay for the storage media as the site grows? Who is on-call when the server goes down?

A final caution must be extended. Always allow the learning goals to set the technical requirements, and never the reverse. It is all too easy, when one has technical expertise, to allow the curriculum to become technocentric. Just because it can be done, doesn't mean it ought to be done! Students have varying levels of access and comfort with technology, but they are open to the argument that they should learn about technology that is common to the business world. Complaints about the inconvenience of going to a computer lab may be handled by analogy with the library: "Your teachers also expect you to go to the library for some resources. This is no different." Above all, students feel rewarded under the pressures of school when you can save them some time. Devising a system where "excellent" work on one assignment excuses students from a later assignment can result in vast improvements in their attitudes and attention to detail.

Perl

Perl is a powerful scripting language that is freely available on almost all computer platforms (www.perl.com). It has its roots in a combination of Unix shell scripts and other common text-processing tools but is now the most widely used language for CGI programming. Although it is an interpreted language, which typically means slower performance, compilers are becoming available. However, it is the ability of Perl in its interpreted mode to create its own code and execute it that makes it useful for individualizing JiTT questions during the delivery process.

Here is an example of an effective and simple CGI Perl script to collect information from students. A typical requirement in a general-education course is to allow students to select a research project, carry that project through in teams, and present it to their classmates. It is often difficult to have students select groups and topics while in class and avoid duplication of ideas. Creating an electronic "bulletin board" to which students post their project proposals allows students to view other's ideas and avoid duplication, while at the same time rewarding early submissions with priority on the project. The teacher may then add comments to each entry simply by editing the HTML file. Making comments to students on the page being viewed by all students allows them to learn from one another's mistakes as well as successes.

The portion of the HTML code that includes the form is shown below, and an example of the page is shown in Fig. 12.3. Note that the full path name for the file to which responses will be appended must be specified within the form. This is accomplished using a hidden <INPUT> tag with NAME equal to FileName and a Value equal to the full path name of the HTML file to which the responses are appended. There is a "magic phrase" inside some hidden text that details where within the original HTML page the proposals will be inserted. Also note the ACTION parameter of the FORM tag in this example. It refers to a CGI called AppendMe.cgi that resides in the cgi-bin folder.

```
<FORM METHOD="POST" ACTION="/cgi-bin/AppendMe.cgi>
Your Names: <INPUT TYPE="text" NAME="student" VALUE=""
   SIZE=15 MAXLENGTH=15 >
<BR><INPUT TYPE="hidden" NAME="FileName"
   VALUE="LabProposals.html">
<TEXTAREA NAME="Proposal" ROWS=10 COLS=80
   WRAP=SOFT></TEXTAREA>
<BR><INPUT TYPE="submit" NAME="action" VALUE="Submit">
</FORM>
<P><HR><P>
<UL>Newest topics are at the top of this list:<BR><BR>
<!- Append Here !->
```

The following script is AppendMe.cgi, the Perl CGI program that receives the submission and posts it back into the page.

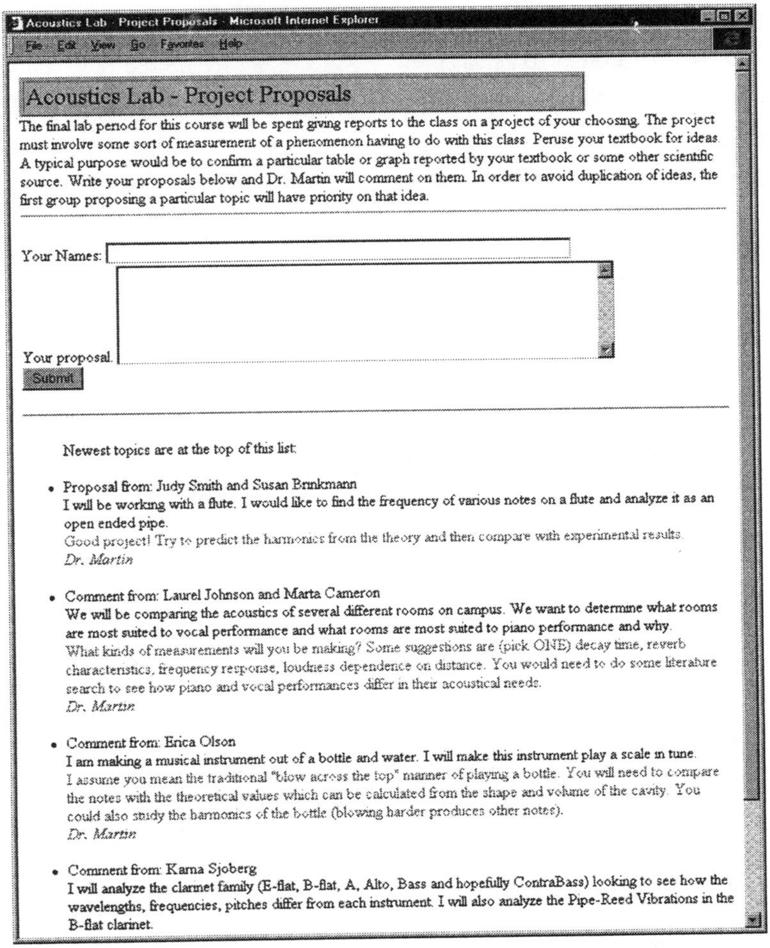

Acoustics Lab - Project Proposals - Microsoft Internet Explorer

File Edit View Go Favorites Help

Acoustics Lab - Project Proposals

The final lab period for this course will be spent giving reports to the class on a project of your choosing. The project must involve some sort of measurement of a phenomenon having to do with this class. Peruse your textbook for ideas. A typical purpose would be to confirm a particular table or graph reported by your textbook or some other scientific source. Write your proposals below and Dr. Martin will comment on them. In order to avoid duplication of ideas, the first group proposing a particular topic will have priority on that idea.

Your Names:

Your proposal:

Submit

Newest topics are at the top of this list:

- Proposal from: Judy Smith and Susan Brinkmann
  I will be working with a flute. I would like to find the frequency of various notes on a flute and analyze it as an open ended pipe.
  Good project! Try to predict the harmonics from the theory and then compare with experimental results.
  Dr. Martin
- Comment from: Laurel Johnson and Marta Cameron
  We will be comparing the acoustics of several different rooms on campus. We want to determine what rooms are most suited to vocal performance and what rooms are most suited to piano performance and why.
  What kinds of measurements will you be making? Some suggestions are (pick ONE) decay time, reverb characteristics, frequency response, loudness dependence on distance. You would need to do some literature search to see how piano and vocal performances differ in their acoustical needs.
  Dr. Martin
- Comment from: Erica Olson
  I am making a musical instrument out of a bottle and water. I will make this instrument play a scale in tune.
  I assume you mean the traditional "blow across the top" manner of playing a bottle. You will need to compare the notes with the theoretical values which can be calculated from the shape and volume of the cavity. You could also study the harmonics of the bottle (blowing harder produces other notes).
  Dr. Martin
- Comment from: Karna Sjoberg
  I will analyze the clarinet family (E-flat, B-flat, A, Alto, Bass and hopefully ContraBass) looking to see how the wavelengths, frequencies, pitches differ from each instrument. I will also analyze the Pipe-Reed Vibrations in the B-flat clarinet.

Figure 12.3 An HTML form that allows students to post project proposals. A CGI both creates the Web page and processes students' responses.

```
#! /usr/local/bin/perl
# AppendMe.cgipl, written by Larry Martin, Nov 14, 1998
# Known bugs:
# some systems may not work well if simultaneous accesses
# are attempted.

use CGI qw(:standard);
use CGI::Carp qw(fatalsToBrowser);
```

```
# For security, allow appends only to files
# within a particular directory that also contains the
# magic phrase.

$file_path='/local/netscape/suitespot/docs/';
$magic_phrase="<!- Append Here !->";

# Get the information from the form
$file_name=param(FileName);
$students=param(Students);
$proposal=param(Proposal);

# Store some other information to help track down
# spoofing

$info = "<!- ".localtime .", ". remote_host()." !->";

# Replace characters that might create security holes in
# some web servers

$proposal =~s/\</&lt;/go;        $proposal =~s/\>/&gt;/go;
$students =~s/\</&lt;/go;        $students =~s/\>/&gt;/go;

# Prepare the item to be inserted
$insertion = "\n<LI>Proposal from: $stu-
   dents\n<BR>$proposal<BR>\n</LI>\n$info<BR>\n";

# Allow slurping in the entire file.
undef $/;
# Open the file and read it in
$append_file = $file_path . $file_name;
open(APPENDFILE,$append_file) or die "Can't find $ap-
   pend_file";
$filecontents=<APPENDFILE>;
close(APPENDFILE);

# Replace the magic phrase with itself along with the
# insertion
$filecontents =~
   s/$magic_phrase/$magic_phrase\n$insertion /;

# Rewrite the file with the new contents
if (open(APPENDFILE,">$append_file")) {
```

```
    print APPENDFILE $filecontents;
    close(APPENDFILE);
} else { die "Insert failed, could not write file.";}

# Redirect the browser to reload the new page
print redirect("/$file_name");

#The End
exit 0;
```

This example includes a Perl module called CGI.pm, which makes it easier to write CGI scripts in Perl, and it is well worth reading its documentation if you are writing CGIs. The latest version of CGI.pm is included in the libraries of the latest distributions of Perl.

A Desktop Level "Starter" System

You can administer CGIs on a Web server running on any platform, including Windows 95, Windows 98, Windows NT, Unix, Linux, or Macintosh. Administering CGIs requires two essential components: a Web server and a programming environment for writing and running CGIs. Web-server software and CGI programming environments abound for all these platforms. We recommend that you select an operating system with which you are already familiar, and a programming environment that you already know or can quickly learn. The accompanying table shows a sample of some Web-server software that you may want to consider. Both commercial products and free software are listed.

Once you've installed your Web-server software, you will probably want to implement JiTT by posting questions to students on the Web and collecting their responses. Collecting and processing student responses is often accomplished using a CGI that runs on the Web-server. Although you have many options for developing and running CGIs, we recommend that you use either Perl or server-side extensions. The advantages of using Perl are: it is fairly easy to learn if you are a programmer; it is available for Macintosh, Windows, and Unix Web servers; and it is powerful, thus allowing you to create very sophisticated CGIs. To download Perl, go to http://www.perl.org/. The version of Perl you download will likely have instructions that help you configure your Web-server to run Perl CGI scripts. For Windows or Unix Web-servers, it is worthwhile to purchase the *Perl Resource Kit for NT* or the *Perl Resource Kit for Unix* from O'Reilly Publishers. The *Perl Resource Kit* greatly simplifies the installation process and provides a complete programming environment for writing and debugging Perl programs. It comes with an editor for writing Perl scripts, a debugger, and a set of help documents and manuals. This kit can greatly improve your efficiency and makes programming in Perl nearly painless.

If you are not a programmer, or if you would simply like to get started with JiTT very quickly, we recommend that you use server-side extensions, which are simply CGIs supplied by the vendor. These extensions, typically accessible via a site-management package such as Microsoft FrontPage or Adobe PageMill, reduce the complexity of writing HTML forms and CGI scripts. The HTML editor that is provided with the site-management package allows you to both create an HTML form and configure the server-side extensions to handle the form information when it is submitted. These extensions are available as drag-and-drop features for easy inclusion in Web pages. In addition, a site-management package may come with predefined templates for creating forms. For instance, a Microsoft FrontPage template includes a survey form that will accept information submitted by a Web browser and will store that information in a text file that you specify. You can use this template to write an HTML form that includes a question and text box for students to type their answers. The answers are then recorded in a text file for you to view before class. Unfortunately, these server-side extensions and site-management packages are not available for all platforms and Web servers. Therefore, you should carefully investigate the requirements of a site-management package and verify that you have the supported Web server and platform.

A Sampling of Commercial and Free Web-Server Software for Various Platforms

Operating System	*Web-Server Software/Publisher*
Macintosh	WebStar/Starnine Personal Web Server/ Microsoft
Windows 95, Windows 98	Personal Web Server: http://www.microsoft.com/windows/ie/pws
Windows NT	Internet Information Server/ Microsoft Personal Web Server/ Microsoft Suitespot/ Netscape
Unix, Linux	Apache/see http://www.apache.org/ Suitespot/ Netscape

Site-Management Packages

Several tools exist that make it easier to create, organize, modify, and maintain Web sites. Examples include Microsoft FrontPage and Adobe SiteMill. Some of these programs are free or included with operating systems or desktop publishing packages. Several features are desirable in such packages, including WYSIWYG HTML editors, syntax checkers, broken-link checkers, predefined CGI programs, and upload utilities, to move a site to another Web server or to duplicate a site.

Although a deep understanding of HTML is often necessary in order to write the most transportable pages (different browsers can give radically different results for the same HTML code), it is often easier simply to write pages within a

WYSIWYG editor and do some fine-tuning at the raw HTML level at the end if necessary. For site-management purposes, it is useful to be able to execute global search and replace functions across several files or even have links updated when documents are renamed or moved. Creating sets of documents off-line and uploading in bulk allows for changes to occur with careful checking before publishing. Local search engines on your Web server can be made more efficient with tracking of keywords placed into META tags in documents. Finally, groups of Web pages may have similar formatting changes made all at once through the use of style sheets, but again, not all browsers take advantage of these. For programming CGIs, there are several Integrated Development Editors that provide easy formatting and even debugging tools.

As described above, you do not have to be a Perl programmer to write CGI scripts for the purpose of JiTT. There are site-management packages that reduce the complexity of writing HTML forms and CGI scripts. These packages usually include two things: server-side extensions for the Web-server you are using and an HTML editor. The HTML editor allows you to both create an HTML form and configure the server-side extension to handle the form information when it is submitted. Also, you can easily embed items like a timestamp or hit-counter that formerly required programming a CGI. Just like forms, timestamps and counters require server-side extensions installed by the site-management package. Site-management packages provide these extensions as drag-and-drop features for easy inclusion in Web pages.

Many-site management packages come with predefined templates for creating your own forms. For instance, Microsoft FrontPage includes a survey form that will accept information submitted by a Web browser and store the information in the specified text file. This is a quick and straightforward way to get started with JiTT, especially if you are uncomfortable with programming. You can write an HTML form that includes a question and text box for students to type their answers. The answers are then recorded in a text file for you to view after the assignment is due.

The disadvantage of using a site-management package is that you are limited to the features it provides. For example, you may need a more complex CGI. In that case, you are advised to modify others' Perl scripts or begin writing your own. However, if your needs are modest, a site-management package may be sufficient for you, and even experienced HTML authors will find a management package useful for routine work before fine-tuning the final code.

Chapter 13: Frequently Asked Questions

The Web

How do you get good images for your Web pages?

Since most Web browsers only support JPEG and GIF image formats, you will need software to acquire and manipulate images in these two formats. GIF images provide the best quality for small icons, buttons, bars, bullets, lines, and grayscale photographs. Compressed JPEGS are best for color photographs. The minor degradation produced by compression is a small price to pay for the greatly reduced image file size.

Learn to use a small number of image-creation and editing tools well. Although many experts will insist on a professional, i.e., expensive, image package, freeware and shareware tools are usually more than adequate. A reasonable starter tool set might consist of the following:

Operating System	Screen Capture	Image Conversion	Image Manipulation
Macintosh	Shift-Cmd-3 (MacOS)	Clip2gif (freeware)Graphic Converter (shareware)	GraphicConverter (shareware) Color-It
Windows	PrintKey (freeware)	GIF Construction Set (shareware), Lview (shareware)	Lview (shareware), Lview Pro

Don't ignore the image-manipulation tools that come bundled with many standard software packages. For example, Microsoft Word and PowerPoint include image-editing capabilities. Although these applications use non-Internet image formats, saving their native documents as HTML files will convert all embedded images into GIF format.

What about copyright problems?

Avoid copyright problems from the start. Media companies, such as news organizations and publishers, are very reluctant to lose copyright since they consider images and other data to be a strategic asset. Use public domain images or make your own using a simple drawing program. Get an inexpensive digital camera (you don't want millions of colors or 1024 x 768 pixel resolution because of file size) and shoot pictures of class demos or public events.

Use applications with which you are already familiar to create graphics. A screen-capture program, such as PrintKey, can be used to grab graphics created with simulation packages such as MathCad or Mathematica. Interactive Physics will not only draw a free-body diagram, it will export the animation as a QuickTime movie. The student version of MicroSim PSpice (free if it is downloaded off of the Web) draws excellent circuit diagrams.

Public domain images are available from a number of sources. Since copyright usually expires after 70 years, there are many more images in the public domain than you might think (we used the original rabbit from Alice in Wonderland in our JiTT logo, for example). Images from government agencies, such as NASA space shots or the excellent photographs of the WPA (Works Progress Administration), are usually in the public domain.

When in doubt, ask the author. Thousands of college and university faculty and students have posted curricular material to the Web, and many are more than happy to have others use their work.

What hardware/software do you need for a Web-server?

Computers are a lot like cars: There isn't a car made that can't be improved with 50 more horsepower. But the family station wagon will get you where you need to go, and many desktop machines will be able to run Web-server software. (It pays to consult an expert if you plan to run extensive CGIs to access student databases. See Chapter 12 by Larry Martin and Aaron Titus for a more detailed discussion).

We have used the following configurations to serve more than a GByte of student and faculty Web pages and to collect form-based JiTT responses at Davidson College (Windows NT) and IUPUI (Macintosh) for the past few years:

Machine Type	Macintosh	Windows NT
CPU	200 MHz PowerPC 604	200 MHz Pentium Pro
Memory	32 MBytes	128 MBytes
Backup	2 GByte JAZ	HP DAT 4 mm
Disk Storage	2 GBytes	3 GBytes SCSI
Server Software	WebStar	Internet Information Server

How do you handle student submissions?

Most of us use a CGI (Common Gateway Interface) written in Perl or in Hypercard. It is also possible to have student submissions directed to a faculty member's e-mail account. Chapter 12 of this book has more details on each of these options.

Do students complain that they don't have access to computers?

No. This is a red herring used by professors and administrators who would rather not learn about computers. Students are, in general, very knowledgeable about consumer uses of the computer, and those who aren't want to learn since they feel they are at a disadvantage without this knowledge. Many students have encountered computers in high school, and a high percentage of students will have computers in their dorm rooms, at home, or at work. Most colleges and universities provide public computer labs, and a few schools either require all students to buy computers or provide computers as part of their tuition package. The hardware required to run browser software and access the Web is minimal compared to the requirements to run a typical office suite or a video game. A sub $1000 computer can, in fact, be used to access all the material in a typical JiTT course.

What if students work together on the Web assignments?

The chance of students working together on Web assignments is no greater than their working together on traditional textbook problems. But one approach is to make working together a positive factor in a student's learning. This can be done in two ways. First, use technology to customize each problem by randomization of key parameters, then encourage students to study together, provided they do not do each other's work. Second, present problems, such as puzzles, that do not have a single answer or that can be solved in multiple ways.

Another approach is to encourage students to work independently on WarmUps by grading them based on their level of effort rather than on correctness (see the next FAQ). Also, when students realize that it benefits them to show what *they* know rather than what their friend knows, because *their* responses will be addressed in the next class session, they will tend to work more independently.

How do you grade the various Web assignments?

Each type of Web assignment is graded with the purpose and function of that assignment in mind. WarmUps are intended to "prime the pump." They generally require the students to do the assigned textbook reading and to then make an honest attempt to connect the textbook words and any previous understanding of the concepts as they formulate their answers. Most JiTT faculty therefore award credit for WarmUps based either on completeness of the answers or on demonstrated level of effort, rather than on correctness. For some types of WarmUps, where questions might be more of a straightforward reading check than a request for synthesis or application, it might be appropriate to grade partially on correctness. However, if the main point of WarmUps is to elicit students' honest answers rather than having the textbook quoted back to the instructor, grading based on correctness can be significantly demotivating and can render the student answers to WarmUps largely ineffective. Since the total amount of credit that can be earned through WarmUps is not very large, grading WarmUps based on completeness or level of effort rather than on correctness does not significantly skew the final point distribution in the course; the vast majority of points are still earned through tests and the final exam.

Puzzles, on the other hand, come at the end of a topic and therefore are tests of understanding. These *are* graded based on correctness. Some faculty grade them as either right or wrong, while others award partial credit for partially correct solutions and/or explanations. Correct solutions offered with no explanation earn no credit. (This serves as an incentive for students to explain their answers and reasoning!)

Answers to the "Good For" questions are also graded based on correctness. Since the answers to the questions are available via the links provided on the "Good For" page, students must find the answer, state it, and give its source in order to earn the credit associated with a given question.

What constitutes a "correct" response to the essay question?

"Correct" can be taken in a couple of different ways. The intention of the essay question (in the "Mindful Interactive Lecture" implementation) is for the students to make a sincere effort to connect the everyday English phrasing of the question, their real-world experience, their prior physics knowledge from any other courses, and their understanding of the textbook reading, and to combine all of that as they answer the question, using proper English sentences. To me, a student who has done that has accomplished what he or she was "supposed" to do, and I consider that to be "correct" (and therefore worth full credit). Of course, this usage of "correct" is based on what the student is supposed to do in responding to the question, not on the correctness of the actual physics content in the answer. A response is technically "correct" if all the connections described above are made *and* the physics content is correctly elucidated. For example, I don't consider an answer that just quotes the textbook to be nearly as good as one that tries to explain in the student's own words what the textbook says. (In fact, in the "Mindful Interactive Lecture" implementation of JiTT, one should avoid asking questions that can be answered by quoting the textbook. Including a phrase such as, "In your own words…" can be very helpful.) Typically, the majority of the responses shown and discussed in class are "correct" in that they have made the connections, but often they lack a full and proper description of the underlying physics.

Do you accept late submissions?

Almost never. The Web-based homework loses *all* of its utility to the instructor and most of its utility to the student if it is completed after the due date, which is shortly before class time. The whole possibility of a feedback loop is destroyed if the submissions come in too late to be read and incorporated into the lesson. Certainly, the student is not as prepared for class if he/she doesn't complete the assigned Web homework prior to coming to class. If a student completes the Web homework after the due date but still before class time, I will accept it (for reduced credit) the first time, and talk with the student about the importance of the due date and time, any extenuating factors that make it difficult for him/her to complete it by the assigned due date, etc.

Do students resent being asked about material before it is covered in class?

This is occasionally an issue with a few students early in the semester, but after only a few lessons, this attitude is almost always dispelled. It is very helpful to explain to the students at the beginning of the course that what is important is that they provide honest answers to the WarmUps and that they will earn credit for completing the assignments, not for the correctness of their answers. Once the instructor explains that he or she will use the students' answers to help maximize the effectiveness of the classroom time together and that the students will help themselves and their fellow students by putting honest effort into their WarmUp answers, the students begin to realize that the WarmUps are intended to help them learn, not to be a "hassle." They start to see that the instructors do not expect them to be able to teach themselves the material perfectly, but that their instructor wants to address the areas most difficult for them. Frequently, in fact, the students end up expressing gratitude that the instructor seems to really care what they do and do not understand and that the WarmUps provide a forum for raising questions to the instructor before class time.

Are the Web assignments a form of homework?

Yes, they are one of several kinds of homework our students complete. Our students still read the textbook, do traditional end-of-the-chapter textbook problems on a frequent, regular basis, and are responsible for completing lab reports and analyses and occasional other paper-based homework, like worksheets or group activities. At the USAFA, we typically assign several textbook questions and problems for each lesson. The JiTT faculty generally have students do the Web-based WarmUps in lieu of the textbook questions so that the total time required for homework remains approximately the same. Just as each of the Web assignments has its own function, each kind of homework serves a unique purpose, and the suite works well to give the students numerous opportunities to learn and practice their problem-solving skills.

How much time should students spend completing the WarmUps?

The total time required to complete a standard WarmUp should be approximately 10–20 minutes. Of course, it is a little hard to define exactly what time is devoted to completing the WarmUp, since the students generally must complete the assigned textbook reading prior to actually answering the questions. Many students may consider the reading time to be time they spend in order to complete the WarmUp, since they might not otherwise do the reading. WarmUp questions should not be so complicated or lengthy that the time required of a student to answer them fully exceeds about 20 minutes. The students should feel that this reasonably short time investment has prepared them for the next lesson and that they've been able to do a reasonable job answering the WarmUp questions in that amount of time. It's important that the time required to complete the WarmUps remains brief, particularly since we do expect our students to be completing other homework for the course.

The Classroom

What do you say about answers that are really bad?

That depends on what "really bad" means. To me, an answer is only "really bad" if it is blank or reflects little or no time investment. If a student fails to make a sincere attempt on a question, he or she will generally explain in the comments box: "I ran out of time to complete this," "My daughter got sick," etc. On the second time or so of a student submitting answers like this, it is a good idea to try to talk with the student (either in person or via e-mail) to express concern and to see if there's anything you (as the faculty member) can do to help.

If "really bad" means "off the wall," "out in left field," or "doesn't have a clue," then what I do depends on how many other "really bad" answers I receive. If there are just a few, I try to address the person individually, either by e-mail or just before or after class. If there is a pattern of such answers from a student, I suggest (in a friendly manner) that they come to my office for a chat about the material. If there are quite a few such responses to a particular question, it's likely that either the question is ill-posed or the students are ill-prepared to respond to it reasonably. In either case, the question and how to approach it should be gone over in considerable detail during the class period. In this situation, it is probably worth mentioning the range of responses received so that the authors of the "really bad" responses don't think they are alone in their difficulty.

Finally, I try to be encouraging and optimistic in responding to student answers. It is usually possible to find *something* positive that can be said about a student's response. Experience with student WarmUp and Puzzle submissions suggests that a lot of improvement in student motivation and attitude is gained through encouraging, positive comments.

When do you discuss the WarmUps, Puzzles, etc?

At IUPUI and at the USAFA, student responses to the WarmUp questions provide the basic framework for the lecture session. As described in detail in Chapter 6, the development of the physics theory follows from a selected subset of the student answers to the WarmUp questions. We make every effort to avoid the perception that the answers to the WarmUps are taken care of separately, either before or after the "lecture."

Students respond to the Puzzle question after they have had considerable exposure to the material (several homework problems and collaborative recitation sessions). The Puzzle question provides the nucleus for the summary session closing a particular topic. We use the opportunity afforded by the analysis of the Puzzle question to teach problem-solving techniques in a formal way.

Do you ever discuss "What is Physics Good For?" in class?

As a rule, "What is Physics Good For?" essays are not discussed in class. They are intended to be enrichment material. All the "Good For" pieces end with questions that the students can answer for extra credit. However, discussing these in class would lead us too far afield. Student answers to the "Good For" questions vary quite a bit, the complexity of the answers being largely a factor of the individual student's interest in the topic treated in the essay. It would be impractical and probably counterproductive to deal with even a small subset of responses.

There are occasions when the "Good For" essay can effectively be integrated into the formal exposition of the material. A good example of such an essay is "The Cassini Mission." The essay deals with the four-year flight to Saturn. NASA has provided a wealth of Web material dealing with all aspects of the physics of the mission. We find that students like space topics such as gravitation and orbital mechanics anyway. Working with a rich resource such as "The Cassini Mission" shifts the focus of the topic away from "Newton's apple" to the scientific and engineering challenges of today's space program. In this way, the "Good For" essay makes a somewhat abstract and apparently remote topic more concrete and interesting. In the current set of about forty essays, we have several examples of this; two are car accident analysis and the Edison vs. Westinghouse fight over the merits of alternating vs. direct current.

Doesn't going over the WarmUps take too much time?

If the WarmUps are regarded as extraneous to the main classroom activity, like the homework, they would be of minimal value and they would take too much time. If, however, they are at the core of what goes on in the classroom, the question disappears.

Preparing a lecture based on WarmUp submissions takes more time than preparing a traditional lecture, especially if you are an experienced instructor not teaching the course for the first time. The minimum requirement is at least skimming the student responses. While a student assistant can categorize and summarize the responses, the lecturer has to incorporate this information into the classroom presentation. In our experience, preparing for a WarmUp-based lecture takes at least 30 minutes. It may more if the lecturer carefully selects the sample responses to take to class and prepares the overhead slides.

How do you cover everything in the syllabus?

The JiTT philosophy subscribes to the less-is-sometimes-more approach. We don't believe that every derivation of every equation has to be explicitly presented to the class by the instructor. We also believe in reducing the number of topics included in the two-semester Introductory Physics Syllabus. For example, we do not cover any modern physics, except peripherally in some topics, such as magnetism, and in the "Good For" essays.

When it comes to covering the material that we do include in our syllabus we prioritize face-time by making the students do some of the time-consuming chores on their own. Students need time to take the concepts apart and put them together again. They proceed at different rates. A rapid didactic delivery may well leave many students behind and give the instructor the illusion that the material has been covered. To take an example from Orbital Mechanics, we require that our students, on their own, work through a very structured Hohmann orbit transfer problem. The problem is presented on the Web at

http://webphysics.iupui.edu/152/GPSLab/gp3ex001.htm

with generous verbal and pictorial hints. Students must work out all the details and submit a written report. This exercise gets very good reviews and, judging from test results, does get the material covered without any expenditure of class time.

Do you do anything special in lab?

In some classes we require that students complete a Web-based pre-lab exercise. The assignment is designed to ensure that students read the preparatory material and work out the relevant physics beforehand. This procedure is superior to a pre-lab in-class quiz, because no lab time is used for the task and the submissions can be examined by the lab instructor before class. At Davidson College, there is a Physlet-based JiTT pre-lab for each lab and sometimes a post-lab JiTT activity as well.

How do you control over-eager students?

This question touches on the difficult issue of managing the active learner classroom, which was discussed in Chapter 5. Try to treat eagerness as a feature rather than a problem. Explicitly state your Q&A policy at the first class meeting (no questions allowed or questions at the end of the period are not among the options). You can make it an announced policy that the instructor will determine when a question is appropriate and in the mainstream and when it is too peripheral. However, no nontrivial question should go unanswered. Peripheral questions and comments can be dealt with in private.

Don't support questions and comments motivated by the urge to show off. When following up on a comment or question from a student, keep the class involved. An extended dialogue between one student and the instructor can do a lot of damage to the atmosphere in the classroom. If the question fits, ask for the class to comment. Good students will often anticipate the next step that you were about to take anyway. Make it a feature.

Do you have an agenda, and do you stick to it?

I have an agenda in the sense that there are certain topics I would like to discuss during each class. I stick to it in the sense that I do not let the students drive the class aimlessly. However, I am flexible about getting to every topic on my list, and I have few preconceived ideas about the order of topics, the emphasis on topics, and the development of the topics.

Managing the classroom discussion is one of the most important aspects of JiTT, and I would suggest rereading Chapters 5 and 6 (or 7) occasionally during your first semester using JiTT. The key notion is to use the student responses to introduce topics and to elicit discussion. You are still in control, and it is still your job to bring up topics, stick to them until conclusions are reached, and put them to rest. The difference is in using students' ideas as cues when you introduce topics and using classroom discussion to monitor when conclusions can be reached.

One classroom activity that I have reduced substantially is my doing derivations. I almost never do the sorts of derivations of the theory that I used to do. I refer students to appropriate sections of the book for that, deriving results on the board only if I believe the treatment in the book is really bad or if students ask questions that can best be answered in the course of a derivation.

Do you have a classroom use for the comments at the bottom of the WarmUp?

As with students' answers to the questions, their comments often make a good starting point to bring up a new topic or review an old one. If one or more students comment that they did not understand a piece of a previous lecture or a section of the text, then you probably should consider discussing that topic in class. When you do so, explain that you are doing so in response to students' requests. This is good for getting students' attention, and it is ideal for improving class morale. When you are overtly responsive to students' needs, they notice, appreciate your effort, and respond in kind.

General

Doesn't this take too much time?

No. It is always tempting to view discussions of the WarmUps and Puzzles as extra topics that must be shoe-horned into an already overloaded lecture session. This is not the case. Rather, it is crucial to remember that these questions are meant to act as cues to the introduction of new ideas (WarmUps) and to review recent topics (Puzzles). We do not suggest that you take a period out of the lecture and use it to "go over" the responses. Instead, read or display one or two sentences out of the responses at a few key points in class when you wish to change topics or elicit discussion.

A related issue, though, is the possibility that you will discover that your students have not understood a topic and that you will have to slow down or back up. This does happen occasionally. If this happens a lot, you will wind up not covering your entire syllabus. Now comes the big question: If your students are not learning everything you cover, wouldn't you rather know this is happening?

How do you deal with students who do not submit answers to the WarmUps?

Send them an e-mail note encouraging their participation and ask them why they are not participating. Listen to their answers. Often, valuable insight can be gained by doing this. We have learned that there are a variety of reasons for nonparticipation, such as technical problems, time pressures, and poor time management/study habits. Some will say they find it useless or that it even inhibits learning. In this case, explain the JiTT philosophy and ask them if they see anything positive that could come from their participation. Explain that several semesters of experience with this strategy at a number of different kinds of institutions suggests that the vast majority of students find it to be helpful.

Also explain that even though they may feel they can earn a very high grade in the course without participating in the Web homework, it is very likely that the amount they actually learn and carry out of the course with them will be increased if they complete the Web homework, which emphasizes connections to the real world.

Do you assign regular homework?

Yes. The amount and format varies among institutions and courses, but we all assign some. In the introductory courses at IUPUI, we have two assignments due each week. They are turned in at the beginning of the recitation section, before we discuss the questions. Each assignment consists of three to five end-of-chapter problems. We only spot-grade these assignments. However, at institutions where graders are available, we would encourage traditional grading.

What are the tests like in this environment?

Again, this varies among JiTT-based courses. However, most of us have not altered our testing strategies from previous years. At IUPUI and the USAFA we give three one-hour exams and a comprehensive final in each course. At IUPUI, the hour exams consist of two multiple-choice conceptual questions worth five or ten points each, followed by three problems in which students must show their work, and for which they may receive partial credit. The final exam is a two-hour test that is similar in form and approximately double in length.

One specific connection between the JiTT aspects of the course and the tests is in motivating students to participate in every facet of the course. We frequently use Puzzle or WarmUp questions (with alterations) for the multiple-choice questions.

Students who do not look at the original assignments soon find out that their fellow students have an advantage. Similarly, we often base one of the three long problems on a problem covered during the collaborative portion of the recitation.

How much weight do you give to the tests?

Most of a student's grade is based on test performance. At IUPUI, each course offers 1000 points of regular credit and about 150 points of extra credit (earned on Puzzles and "What is Physics Good For?" assignments). Of this, the hour exams offer 300 points, the final offers 300, the homeworks and WarmUps combine to 250, and the lab offers 150. Since the homework and lab points are relatively easy to earn, the tests account for almost all of the distinctions among students. In order to assure that students do not pass the course on the strength of homework, extra credit, etc., we require a minimum of 50% on the tests alone for a passing grade.

Do you use the Web for any traditional assessment?

Only to the extent that the WarmUps are graded. The weight of the Web work varies by institution and course. At IUPUI, the WarmUp points represent about 10% of the students' grades. At the USAFA, the Web component can be up to about 10% of the total points but is typically more like 5%. We know that the students may have received help with a Web submission; however, the same has always been true of written homework. We assign homework and WarmUps for regular credit, and we spot-grade students' efforts. Most of the assignment of final grades is determined by student performance on the hour exams and final exam.

How well does JiTT work?

We have been developing and employing the essential elements of the JiTT strategy in our physics courses over the past five semesters and are convinced that JiTT makes a significant difference in what our students take away from their courses. The JiTT strategy is designed to encourage both attitudinal and cognitive gains. Early assessment data, coupled with considerable anecdotal evidence, suggest that JiTT is indeed effective in these areas.

Thus far, our assessment efforts have been directed into three broad areas: course attrition rates, student cognitive gains, and student attitudes. In the attrition/withdrawal area, we are comparing course attrition rates prior to and since the inception of JiTT-based courses at IUPUI. Attrition at IUPUI has dropped by approximately 40% in each of two courses since the inception of JiTT-based courses in the introductory sequence. In the cognitive gain area, we are in the early stages of a GPA comparison of JiTT versus "no JiTT" sections of introductory physics at the USAFA. We are also considering using other metrics, such as the Force Concept Inventory test, to assess JiTT cognitive effectiveness. A controlled comparison of JiTT/"no JiTT" sections of the second-semester introductory course at the USAFA is planned for the Spring 1999 semester (after this book goes to press). In the student attitude area, we have been asking our students to complete anonymous end-of-semester surveys about their opinions and thoughts about their course in general and JiTT in particular. The data we have acquired from these surveys at IUPUI, the USAFA, and Davidson are overwhelmingly positive. We have also begun offering the Maryland Physics Expectation (MPEX) survey developed by Jeff Saul and others at the University of Maryland Physics Education Research Group. After more MPEX data are obtained, we will compare student attitudes toward physics before and after their JiTT-based courses and will compare the JiTT and "no JiTT" students' attitudes.

We are continually acquiring more assessment data and analyzing these data. For current assessment and evaluation information, please visit the JiTT Web site:

http://webphysics.iupui.edu/jitt.html.

As a final point about JiTT effectiveness, it is worth noting that numerous faculty from a diverse set of institutions across the country have taken JiTT workshops or heard JiTT talks and implemented JiTT in their courses as a result. Current information about these other JiTT adopters and their courses is also available at the JiTT Web site.

References

AIP Education and Employment Statistics Division. (1995) "Skills Used Frequently by Physics Bachelors in Selected Employment Sectors."

Britton, Bruce K. and Abraham Tesser. (1991) "Effects of time-management practices on college grades," *Journal of Educational Psychology* **83**, 405–410.

Cashin, W. E. (1985) "Improving Lectures," Idea Paper No. 14, Manhattan: Kansas State University Center for Faculty Evaluation and Development.

Christian, Wolfgang and Titus, Aaron. (1998) "Developing Web-Based Curricula Using Java Applets," *Computers in Physics,* **12**, 227–232.

Cope, R. & Hannah, W. (1975) *Revolving College Door: The Causes and Consequences of Dropping Out and Transferring*, Wiley, New York.

Devlin, K. (1998) "Rather than scientific literacy, colleges should teach scientific awareness," *Am. J. Phys.*, **66**, 559–560.

Fuller, Robert G. (1982) "Solving physics problems—how do we do it?" *Physics Today*, September 1982, 43–47.

Hake, Richard R. (1987) "Promoting student crossover to the Newtonian world," *Am. J. Phys.*, **55**, 878–884.

Hake, Richard R. (1991) "My Conversion to the Arons-Advocated Method of Science Education" *Teach. Educ.*, **3** (2), 109–111.

Hake, Richard R. (1998) "Interactive-engagement vs. traditional methods: A six-thousand-student survey of mechanics test data for introductory physics courses," *Am. J. Phys.*, **66**, 64–74.

Hestenes, David. (1998) "Guest Comment," *Am. J. Phys.*, **66**, 465–467.

Halloun and Hestenes. (1985) "The Initial knowledge state of college physics students," *Am. J. Phys.*, **53**, 1043–1055.

Heller P., Keith R., and Anderson S. (1992) "Teaching problem solving through cooperative grouping. 1. Group vs. individual problem solving," *Am. J. Phys.*, **60**, pp. 627–636.

Heller P. (1992) and Hollabaugh M. "Teaching problem solving through cooperative grouping. 2. Designing problems and structuring groups," *Am J. Phys.*, **60**, 637–644.

Jonassen, David H. and Grabowski, Barbara L. (1993) *Handbook of Individual Differences, Learning, and Instruction*, Lawrence Erlbaum Associates, New Jersey:

Karplus, Robert. (1977) *J. Res. Sci. Teach.*, **14**, 169.

Langer, Ellen J. (1997) *The Power of Mindful Learning*, Addison-Wesley, Reading, MA.

Laws, Priscilla. (1991) "Calculus-Based Physics Without Lectures," *Phys. Today* **44** (12), 24–31.

Laws, Priscilla. (1997) "Millikan Lecture 1996: Promoting active learning based on physics education research in introductory physics courses," *Am. J. Phys.*, **65**, 13–21.

Malone, T. (1981) "Toward a Theory of Intrinsically Motivating Instruction," *Cognitive Science,* **4**, 333.

Mazur, E. (1996) *Peer Instruction: A User's Manual*, Prentice-Hall, Upper Saddle River, NJ.

McDermott, L. (1991) "Millikan Lecture 1990: What we teach and what is learned—Closing the gap," *Am. J. Phys.*, **59** (4), 301–315.

McDermott, Lillian C. and Shaffer, Peter S. (1998) *Tutorials in Introductory Physics*, Prentice-Hall, Upper Saddle River, NJ.

McKeachie, W. J., Pintrich, P. R., Yi-Guang, L., and Smith, D. A. F. (1986) "Teaching and Learning in the College Classroom: A Review of the Research Literature," Ann Arbor: Regents of the University of Michigan.

Monaghan, P. (1998) "U. of Washington Professors Denounce Governor's Embrace of On-Line Education," *The Chronicle of Higher Education*, http://chronicle.com/free/98/06/98060801t.shtml.

Peters, P.C. (1992) "Even honors students have conceptual difficulties with physics," *Am. J. Phys.*, **50** (6), 501–508.

Pew Higher Education Roundtable and the Knight Collaborative. (1998) "A Teachable Moment," Policy Perspectives, **8** (1).

Redish, E. F. (1994) "Implications of cognitive studies for teaching physics," *Am. J. Phys.*, **62** (9), 796–803.

Sutherland, Tracey E. and Bonwell, Charles C. eds., (1996) *Using active learning in college classes: a range of options for faculty*, San Francisco: Jossey-Bass,

Titus, Aaron P. (1998) "Integrating Video and Animation with Problem Solving Exercises on the World Wide Web," Ph.D. diss., North Carolina State University.

Titus, Aaron P., Martin, L. W. and Beichner, Robert J. (1998) "Web-based Testing in Physics Education: Methods and Opportunities," *Computers in Physics*, **12** (2) 117–123.

Tobias, Sheila. (1990) *They're Not Dumb, They're Different: Stalking the Second Tier*, Tucson, AZ: Research Corporation.

University of Maryland Fermi Problems Web site: http://physics.umd.edu/rgroups/ripe/perg/fermi.html

Van Heuvelen, Alan. (1991) "Overview, Case Study Physics" *Am. J. Phys.*, **49**, 898–907.

Index